Edible *and* Medicinal FLORA of the West Coast

THE PACIFIC NORTHWEST *AND* BRITISH COLUMBIA

UNIVERSITY OF WASHINGTON PRESS
Seattle

UNIVERSITY OF WASHINGTON PRESS
uwapress.uw.edu

Published simultaneously in Canada by by Heritage House Publishing Company Ltd. heritagehouse.ca.

LIBRARY OF CONGRESS CATALOGING-IN-PUBLICATION DATA ON FILE

ISBN 978-0-295-74804-7 (paperback)

Edited by Warren Layberry
Proofread by Renate Preuss
Cover and interior book design by Jacqui Thomas
Cover photographs: counter-clockwise from top right, Bunchberry, Shaggy Mane, Spreading Stonecrop, Swamp Rose, and Blue-berried Elder by author; and Sea Palm by Ron Wolf
All interior photos are by Collin Varner, except as follows under the Creative Commons Attribution 2.0 license (creativecommons.org/licenses/by/2.0/): California hazelnut, p. 170, Superior National Forest, USFS, Flickr; Curled dock, p. 44 (both), Harry Rose, Flickr; Marshpepper smartweed, p. 75 (left), K. Kendall, Flickr; Sugar wrack, p. 224, J Brew. All images have been altered to fit the layout.

The interior of this book was produced on FSC®-certified, acid-free paper, processed chlorine free, and printed with vegetable-based inks.

24 23 22 21 20 1 2 3 4 5

Printed in China

Edible *and* Medicinal FLORA of the West Coast

CONTENTS

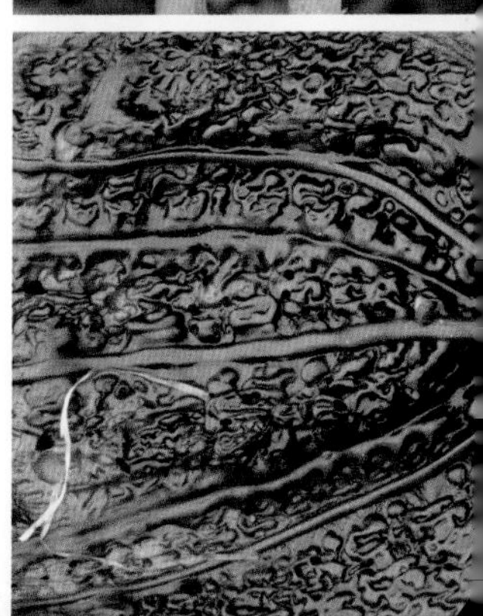

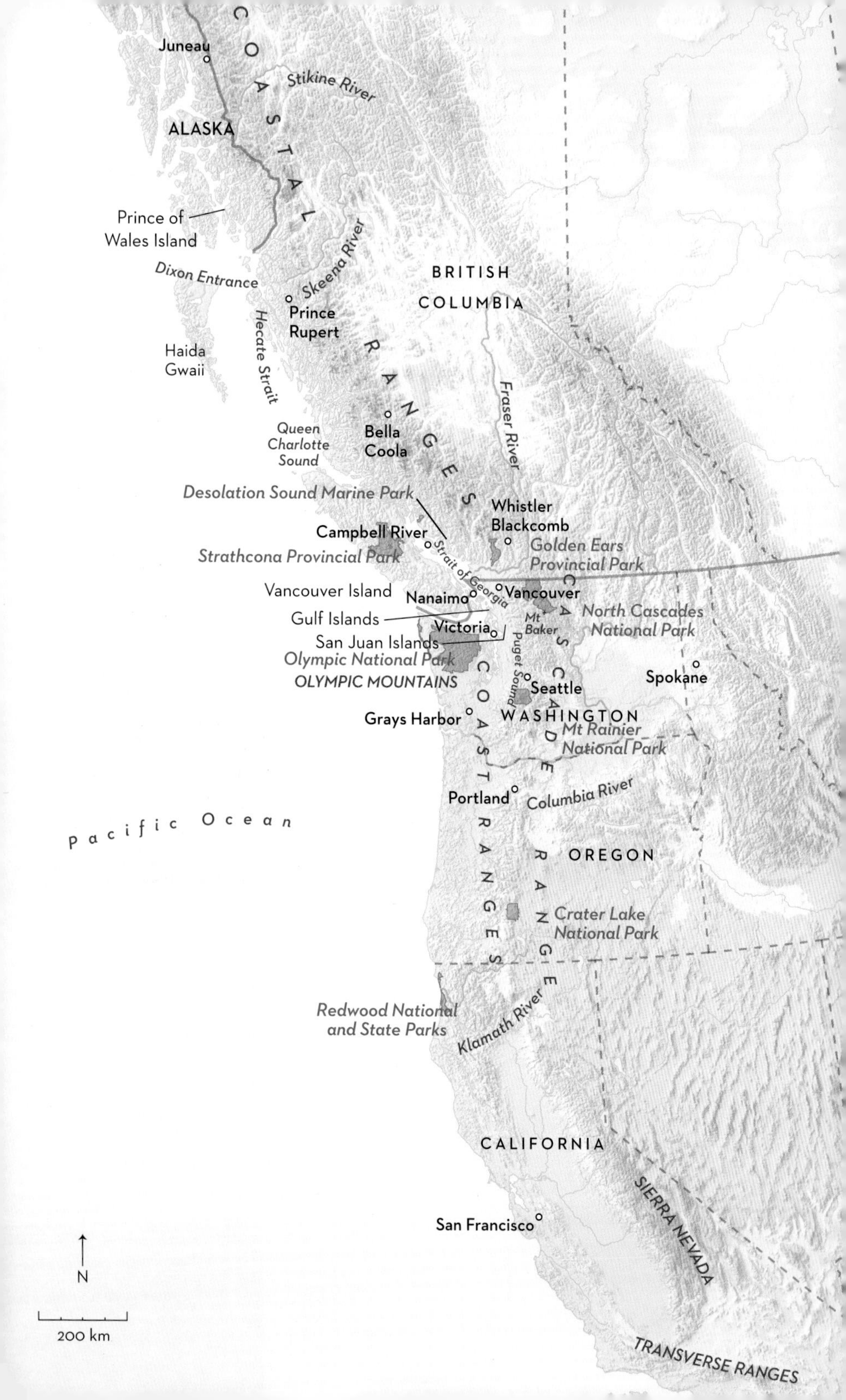
Juneau
COASTAL RANGES
Stikine River
ALASKA
Prince of Wales Island
Dixon Entrance
Skeena River
Prince Rupert
BRITISH COLUMBIA
Haida Gwaii
Hecate Strait
Fraser River
Queen Charlotte Sound
Bella Coola
Desolation Sound Marine Park
Whistler Blackcomb
Campbell River
Golden Ears Provincial Park
Strathcona Provincial Park
Strait of Georgia
Vancouver Island
Nanaimo
Vancouver
Gulf Islands
North Cascades National Park
Mt Baker
Victoria
San Juan Islands
Puget Sound
Olympic National Park
OLYMPIC MOUNTAINS
Spokane
Seattle
CASCADE RANGE
Grays Harbor
WASHINGTON
Mt Rainier National Park
COAST RANGES
Portland
Columbia River
Pacific Ocean
OREGON
Crater Lake National Park
Redwood National and State Parks
Klamath River
CALIFORNIA
SIERRA NEVADA
San Francisco
N
200 km
TRANSVERSE RANGES

INTRODUCTION

THIS BOOK DESCRIBES OVER 200 edible plants and fungi, some native and some introduced, that can be found in the coastal Pacific Northwest. It is not intended to present an exhaustive list of edible species within the region, but rather to provide an overview of various common and not-so-common species representative of the Pacific Northwest.

While some of the species listed also have long traditions of medicinal use, the sole criteria for inclusion in this book is edibility, meaning that the "medicinal" species listed are but a subset of the edible species selected.

The vast majority of species included in this book were photographed by the author on his exploration of the coastal region over a period of several years. Written with the average observer in mind, the book has an intentional bias toward plants and fungi that are visible to the typical viewer rather than more obscure hard-to-find species. However, it is also designed to be a quick, useful resource for readers of all levels of expertise.

For the purposes of this book, the Pacific Northwest is defined as the region stretching from Juneau, Alaska, to San Francisco, California, from the mainland coast to approximately 100 kilometres (60 miles) inland, as well as the various coastal islands.

Despite the Pacific Northwest being a very large and diverse region, the species within it are relatively uniform. Yet, venture another 50 kilometres (30 miles) inland, and the ecosystem is vastly different. The species covered by this book are the delights that the ambler encounters, from the intertidal to the subalpine areas.

▶ Disclaimer

While all the species listed are at least marginally edible (or have at least one edible component), it is important to recognize that edibility comes with a great many caveats, and care must be taken if you intend eat anything you find in the wild. Some species have both edible and poisonous components. Some species are edible only when prepared in a certain way or collected at a certain time of year. Some edible species are hard to distinguish from poisonous related species. Furthermore, edible doesn't necessarily mean palatable.

Although this book mentions some of the traditional uses of edible plants for medicinal or other purposes, the author does not advocate the use of these plants for such purposes. Many plants in this region, including those with some medicinal uses, can be harmful or even poisonous. The author and publisher do not recommend experimentation with plants found in the wild and are not responsible for misidentification by readers of any species found in this book.

▶ How to use this book

This book was written to be accessible to anyone, from amateur enthusiasts to professionals. Gardeners and nature lovers will enjoy discovering the rich sampling of edible flora that can be found right in their own backyard. Hikers and campers will delight in identifying plants encountered on the trail and in the forest. And academics and those working in the field will find a useful, general overview of edible species within the Pacific Northwest as a whole.

Keep this book in your car, or put it in your backpack for a day's hike or an extended camping trip so you can look up the species you find and learn more about them. Use it to engage with your surroundings and deepen your understanding of the region and the environment.

This increased knowledge will add to your enjoyment of being outdoors, bring you closer to nature, and help you get more out of your experiences. Familiarize yourself with the species that can be found within the Pacific Northwest, and then see what you can find.

▶ What you will find inside

This book is divided into six sections. They are **Flowering Plants**; **Berries**; **Ferns**; **Trees, Shrubs, and Bushes**; **Fungi and Allies**; and **Marine Plants**. Within each section the species are listed alphabetically by common name. For each species, you will find a general description, including dimensions, colouring, and other points that may be useful in finding or identifying the particular species. Also included is information on where the species can be found, in terms of both its preferred habitat and its specific location.

Etymologies of common names, genus names, and species names have also been provided where they may be of interest. Species with traditional uses—medicinal or otherwise—will list them.

All species will list the nature and limits of their edibility, including traditional use or importance. Also listed will be important cautions as to potential hazards of consuming, or in some cases even handling, the species in question.

NOTE: When considering the consumption of any wild species of plant or fungus, caution and discretion are your friends. If a potential identification does not agree with some aspect of the listing, better to leave the specimen where it is.

▶ The varied environment of the Pacific Northwest

The Pacific Northwest comprises thousands of kilometres of coastline, including an extensive mainland coast and numerous inlets and coastal islands. For an overview of the region, see the map on page *vi*. Notable islands include Prince of Wales Island, off the coast of Alaska; Vancouver Island and Haida Gwaii, both part of British Columbia; and the San Juan Islands, in Washington.

It is at the ocean's edge that thousands of rivers and streams, which can be thousands of kilometres long, finally empty their fresh water into the ocean. With the fresh water comes silt. Some of the silt accumulates, forming small islands within the rivers. However, the majority gets deposited at the mouths of the rivers as nutrient-rich deltas. Some of these deltas are so large that they support thousands of hectares (or acres) of farmland, producing some of the finest crops in North America.

To most people, Mount Rainier, Mount Baker, and the peaks of the Whistler Blackcomb region conjure up images of wonderful snowfalls and winter sports. But these mountains are quickly expanding as all-season destinations with endless opportunities for outdoor recreation, such as hiking, cycling, canoeing, and observing nature. This book looks at our mountains when they are dressed in green, red, yellow, blue, and all the colours in between.

The environment of the Pacific Northwest is influenced by forces from all directions. From the east, inland rivers bring nutrients from the mainland; from the west, the coast is shaped by the ocean tides and winds, as well as human activities such as shipping and boating. Species introduced accidentally, most often from Europe and Asia, tend to thrive in the Pacific Northwest due to its temperate climate and high levels of precipitation. Interestingly, the reverse does not also hold true—species carried west across the Pacific Ocean do not tend to take hold in other areas.

GLOSSARY

anther	The pollen-bearing (top) portion of the stamen
axil	The angle made between a stalk and the stem on which it is growing
biennial	A plant that completes its life cycle in two growing seasons
bract	A modified leaf below the flower
catkin	A spike-like or drooping flower cluster, either male or female, found in species such as cottonwood
coniferous	Having evergreen leaves, usually needle-like or scaled
corm	A swollen underground stem capable of producing roots from its base plate, and leaves and flowers from its tip—similar to a bulb, but flatter and less egg-shaped, and lacking the outer sheathing
deciduous	Having parts (leaves, bark) that shed annually, usually in the autumn
dioecious	Having male and female flowers on separate plants
epiphyte	A plant that grows on another plant for physical support without robbing the host of nutrients
fibrous-rooted	Having no central taproot but instead many thin shallow roots—grasses, for instance, are fibrous-rooted
herbaceous	A non-woody plant that dies back into the ground every year
hip	A fruit typically associated with rose species; the rose hip contains the seeds

holdfast	A hard root-like structure used by seaweeds to attach themselves to rocks and the ocean floor
lanceolate	Tapering to a point at the tip and sometimes at the base
node	The place on a stem where the leaves and axillary buds are attached
obovate	An oval-shaped leaf, with the narrower end pointing downward, like an upside-down egg
panicle	A branched inflorescence, or cluster of flowers
perennial	General term for a plant that lives above the ground throughout the year
petiole	The stalk of a leaf
pinnate	A compound leaf with the leaflets arranged on both sides of a central axis
rhizome	An underground modified stem, generally shallow-rooted and growing horizontally; it can produce new plants close to the parent plant
sepal	The outer parts of a flower; usually green
sori	Spore cases
stipe	A stem or stalk, such as on a mushroom, bull kelp, or the maidenhair fern
stolon	A stem or branch that runs along the surface of the ground and takes root at the nodes or apex, forming new plants
style	The stem of the pistil (female organ)
taprooted	Having a central root that grows downwards with thin lateral roots branching off, as with carrots and dandelions

FLOWERING
PLANTS

ALFALFA LUCERNE / BUFFALO HERB / PURPLE MEDIC / MU-SU / JATT / YONCA

Medicago sativa

PEA FAMILY Fabaceae

Description Alfalfa is a deeply taprooted herbaceous perennial that grows to 90 cm (36 in.) in height. Its flowers are bluish purple and borne in terminal clusters arising from the leaf axils. The seed pods are brown and spirally coiled at maturity. The leaves are alternating, hairy, and pinnately compounded into three leaflets.

Origin Introduced from Eurasia as a forage crop.

Etymology The genus name *Medicago* refers to Media, Ancient Persia, where the plant is thought to have first grown. The common name alfalfa is derived from the Arabic word *alfalfas*, which means "father of all foods."

Habitat Usually seen as a pasture crop at low to moderate elevations.

Season Flowers from May to September.

Traditional/Medicinal Use The Chinese have been using alfalfa since the sixth century to treat kidney stones and to reduce fluid retention and swelling. Early American settlers used it to treat scurvy, cancer, boils, bedsores, and urinary and bowel problems.

Edibility Alfalfa sprouts have in recent times become popular in salads and sandwiches.

ALPINE FIREWEED BROAD-LEAVED FIREWEED

Epilobium latifolium

EVENING PRIMROSE FAMILY Onagraceae

Description Alpine fireweed is a showy herbaceous perennial that grows to 40 cm (16 in.) in height. The large flowers are rose to purple in colour and contrast well with the lanceolate bluish-green leaves.

Habitat Gravel bars along streams and creeks or on wet slopes at mid to high elevations.

Season Flowers in July at mid elevations and in August at higher elevations.

Edibility The new shoots are edible and can be used as a potherb, as can young leaves, while flowers can be used to brighten up a salad.

AMERICAN BROOKLIME

Veronica americana

FIGWORT FAMILY Scrophulariaceae

Description American brooklime is an herbaceous perennial with flowering stems up to 1 m (3.3 ft.) tall. The light blue–mauve flowers are produced in loose clusters and have two obvious stamens. The leaves are opposite, usually with three to five pairs per flowering stem, and oval to lance shaped.

Habitat Ditches, ponds, and marsh edges at low to mid elevations.

Season Flowers from July to September.

Traditional/Medicinal Use Brooklime was once considered an important medicine in Europe and among North American Indigenous groups, up and was used to treat asthma and bronchitis. Once considered an antiscorbutic, it was used to help ward off scurvy.

Edibility The leaves are edible and can be used as a potherb or raw in salads.

AMERICAN POKEWEED

COMMON POKEBERRY

Phytolacca americana

POKEWEED FAMILY Phytolaccaceae

Description Pokeweed is an herbaceous perennial up to 3 m (10 ft.) in height. Its flowers are white, five petalled, and borne in columnar clusters. The resulting fruit start off green and end a beautiful dark purple. The leaves are up to 15 cm (6 in.) long, alternating, and lanceolate.

Origin Native to eastern North America.

Etymology The genus name *Phytolacca* is from the Greek words *phyton*, for "plant," and *lac*, referring to the dye extracted from the lac insect. The common names pokeweed and pokeberry are derived from *pocan*, an Indigenous word for a plant that yields a dye. Indigenous groups made a red dye from the berries to colour baskets and paint their horses.

Habitat Very common along roadsides, pastures, and forest openings and edges.

Season Flowers from July to August, with the berries ripening from September to October.

Traditional/Medicinal Use Pokeweed has a long history among Indigenous groups. The roots are considered to be a blood cleanser, and capsules are sold in modern-day herbal shops. Pokeweed was being studied as a hopeful anti-AIDS drug and as a cure for childhood leukemia.

Edibility The young shoots and leaves can be boiled twice to remove any toxins and eaten like asparagus or spinach; however, the leaves become toxic as they age.

CAUTION The raw berries are poisonous.

ANNUAL SOW THISTLE

PRICKLY SOW THISTLE / SPINY MILK THISTLE

Sonchus asper

ASTER FAMILY Asteraceae

Description Annual sow thistle is a taprooted annual that grows to 1.2 m (4 ft.) in height. Its yellow flowers are up to 2.5 cm (1 in.) across. It has ray florets only. The floral bracts have no hair and are borne in flat-topped clusters. The leaves are the identifying factor when distinguishing annual sow thistle from common sow thistle. Both species have spiny-toothed margins on the leaves, but the base of annual sow thistle's leaves have large rounded flanges as they clasp the stem. With common sow thistle, the base of the leaves is pointed as the leaves pass through the stem.

Origin Introduced from Europe.

Etymology As the common name sow thistle suggests, it is a favourite food for pigs. The genus name *Sonchus* is derived from the Greek word *sonchos*, meaning "hollow," referring to the hollow stems.

Habitat Waste places, forest openings, pastures, and gardens.

Season Flowers throughout the summer, from June to September.

Traditional Use The milky juice of all sow thistles is said to be a healthy wash for the skin.

Edibility Like other sow thistles, the young leaves are excellent in soups, casseroles, and salads.

BITTER DOCK BUTTER DOCK / BROAD-LEAVED DOCK

Rumex obtusifolius

BUCKWHEAT FAMILY Polygonaceae

Description Bitter dock is a large taprooted herbaceous perennial up to 1.2 m (4 ft.) in height. Its tiny flowers are a plain greenish brown and borne in dense elongated clusters. The basal leaves are up to 30 cm (12 in.) long, alternating, heart shaped at the base, and reduced in size upward.

Similar Species Clustered dock (*R. conglomeratus*) is very similar. It can be distinguished by its segmented flower clusters.

Origin Introduced from Eurasia.

Etymology Until modern times, the large leaves were used to wrap bricks of butter; hence the common name butter dock. The species name *obtusifolius* means "blunt leaved."

Habitat Can be extremely invasive in moist lowland pastures and areas with nutrient-rich soils.

Season Flowers from June through summer.

Traditional/Medicinal Use Medicinally, the leaves were applied to burns, scalds, and nettle stings.

Edibility The young leaves can be used as potherbs, and the seeds can be dried, ground, and added to flour.

BLACK LILY RICE ROOT

Fritillaria camschatcensis

LILY FAMILY Liliaceae

Description Black lily is an herbaceous perennial up to 60 cm (24 in.) in height. Its nodding flowers have six petals and are bell shaped, purple brown, and up to 3 cm (1 in.) across. The leaves are in whorls, lance shaped, and grow to 8 cm (3 in.) long. Chocolate lily (*F. lanceolata*) is a less sturdy species with thinner leaves and mottled flowers.

Etymology The common name rice root comes from the large white bulbs, which are covered with rice-like scales. The genus name *Fritillaria* refers to the flower's checkered pattern, reminiscent of old dice boxes.

Habitat Open forests and moist grassy fields at low to high elevations.

Season Flowering starts mid-May.

Edibility The bulbs can be boiled or steamed. Black lily bulbs were eaten by most Pacific Northwest peoples.

BRITTLE PRICKLY-PEAR CACTUS

Opuntia fragilis

CACTUS FAMILY Cactaceae

Description Brittle prickly-pear cactus is a mat-forming, well-armed perennial up to 60 cm (24 in.) across. Its yellow flowers are tissue-like and up to 5 cm (2 in.) across. The leaves are modified succulent stems, each carrying very pointed spines up to 3 cm (1 in.) long.

Habitat Dry exposed sites with well-drained soil.

Season Flowers from June to July.

Traditional Use The sturdy, sharp spines were used to pierce ears.

Edibility After the spines are carefully removed, the fleshy sections can be roasted and eaten directly or added to soups and stews. Our prickly-pear cacti are smaller than the southern ones; however, they were used as food wherever they grew.

BROAD-LEAVED STONECROP

Sedum spathulifolium

STONECROP FAMILY Crassulaceae

Description Broad-leaved stonecrop is a rhizomatous spreading succulent up to 8 cm (3 in.) tall. The spoon-shaped leaves are flattened more than those of most sedums. The attractive yellow flowers are formed in flat-topped clusters, which contrast well with the green to reddish leaves.

Etymology Stonecrop's old meaning was to crop or remove the leaves or plants from stone.

Habitat Mainly found on coastal bluffs to mid elevations.

Season Flowers from June to August.

Edibility The leaves can be consumed raw (in moderation) and were eaten by Haida people.

BUNCHBERRY DWARF DOGWOOD

Cornus canadensis

DOGWOOD FAMILY Cornaceae

Description Bunchberry, a perennial no higher than 20 cm (8 in.) tall, is a reduced version of the Pacific dogwood tree (*C. nuttallii*). The tiny greenish flowers are surrounded by four showy white bracts, just like the flowers of the larger dogwood. The evergreen leaves, 4–7 cm (2–3 in.) long, grow in whorls of five to seven and have parallel veins like the larger tree. The beautiful red berries form in bunches (hence the name) just above the leaves in August.

Habitat From low to high elevations in cool, moist, coniferous forests and bogs.

Season Bunchberry and Pacific dogwood have a habit of flowering twice, once in the spring and then again in the late summer.

Edibility Bunchberries can be eaten raw. Coastal Indigenous groups often ate them with grease in late summer to early autumn.

CALIFORNIA POPPY

Eschscholzia californica

POPPY FAMILY Papaveraceae

Description California poppy is an herbaceous perennial up to 50 cm (20 in.) tall with a long taproot. The flowers are electric golden orange, up to 5 cm (2 in.) across and borne singly on low stalks. The dull green fern-like leaves contrast well with the flowers.

Etymology California poppy is California's state flower. The botanist Adelbert von Chamisso (1781–1838) named the California poppy genus for Johann Friedrich von Eschscholtz (1793–1831), the physician on the Russian ship *Rurik* that landed in San Francisco Bay in 1816, where he found the plants.

Habitat Dry rocky soils. In Washington and southern BC, the California poppy is considered a garden escapee.

Season Flowers throughout the summer, depending on when the seeds germinated.

Edibility The electric orange flowers are edible and make a colourful addition to a salad.

CANADA GOLDENROD

Solidago canadensis

ASTER FAMILY Asteraceae

Description Canada goldenrod is an herbaceous perennial of various heights, from 0.3 m (1 ft.) to 1.5 m (5 ft.). Its small golden flowers are densely packed to form terminal pyramidal clusters. The many small leaves grow at the base of the flowers; they are alternate, lance linear, and sharply saw-toothed to smooth.

Habitat Roadsides, wastelands, and forest edges at low to mid elevations.

Season Flowers from July to August.

Edibility The fresh leaves can be used as a potherb and the dried leaves for tea.

▶ CATTAIL

Typha latifolia

CATTAIL FAMILY Typhaceae

Description Cattails are semi-aquatic perennials that can grow to 2.5 m (8 ft.) in height. The distinctive "tail," a brown spike, is 15–20 cm (6–8 in.) long, 3 cm (1 in.) wide, and made up of male and female flowers. The lighter-coloured male flowers grow at the top and usually fall off, leaving a bare spike above the familiar brown female flowers. The sword-shaped leaves are alternate and spongy at the base.

Habitat Common at low to mid elevations, at lakesides and riversides, and in ponds, marshes, and ditches.

Season Seed heads can be seen from July to December.

Traditional Use The long leaves were used to weave mats, and the fluffy seeds were used to stuff pillows and mattresses.

Edibility The lower parts of the leaves and new shoots can be eaten raw (added to salads) or roasted. Even the yellow pollen can be sprinkled on bread or cheese.

CHICORY

Cichorium intybus

ASTER FAMILY Asteraceae

Description Chicory is an attractive perennial that grows to 1.9 m (6.2 ft.) tall. The beautiful blue flowers are up to 5 cm (2 in.) across, with each ray having five teeth at the tips. The coarse-looking leaves get progressively smaller from the base of the stem to the tips of the branches.

Origin Chicory is native to the Mediterranean.

Habitat Being drought resistant, it grows well on roadsides, abandoned fields, pastures, bare land, and ditches.

Season Flowers from mid- to late summer.

Edibility The young leaves can be used in salads, and the roots baked, ground, and used as a coffee substitute.

CHOCOLATE LILY

Fritillaria lanceolata

LILY FAMILY Liliaceae

Description Chocolate lily is an herbaceous perennial from bulb up to 80 cm (32 in.) in height. Its nodding flowers are dark brownish purple with greenish-yellow mottling. Each bell-shaped flower has six petals up to 3 cm (1 in.) across. The leaves are lanceolate and formed in one or two whorls of 2–5 leaves. This is one of the Pacific coast's most prized spring flowers.

Etymology The genus name *Fritillaria* refers to the flower's checkered pattern, reminiscent of old dice boxes.

Habitat Exposed grassy bluffs and meadows and dry to moist soil at low to mid elevations.

Season Flowers from April to May.

Edibility The bulbs can be boiled or steamed. Chocolate lily bulbs were eaten by most Pacific Northwest peoples.

CLEAVER BEDSTRAW

Galium aparine

MADDER FAMILY Rubiaceae

Description Cleaver is a sprawling, clinging, or climbing annual up to 60 cm (24 in.) long. Its small white flowers are stalked from the leaf axils. The resulting fruit are annoying little burrs covered with hooked bristles. The bristly leaves are very narrow and grow to 5 cm (2 in.) long in whorls of six to eight. The back-angled bristles on the square stems and leaves help the plants climb over and through other plants.

Habitat Most commonly seen on or near beaches climbing over rocks and logs, in disturbed sites, and in broad-leaved forests.

Season Flowers in the spring, with the burrs maturing in July.

Traditional Use The abrasive parts of the plant were rubbed between the hands to remove pitch.

Edibility A caffeine-free coffee substitute can be made from the dried and roasted nutlets.

▶ COMFREY KNITBONE / BONESET

Symphytum officinale

BORAGE FAMILY Boraginaceae

Description Comfrey is a taprooted perennial 0.9–1.5 m (3–5 ft.) in height. Its bell-shaped flowers are creamy yellow to blue or purple and droop in one-sided clusters, much like a scorpion's tail. The leaves are alternating, hairy, and rough in texture and can be over 30 cm (1 ft.) long at the base.

Origin Introduced from Europe.

Etymology The Greek genus name *Symphytum* as well as the common names comfrey, knitbone, and boneset refer to this plant's usefulness in mending broken bones.

Habitat Moist areas, meadows, ditches, waste areas, and roadsides.

Season Flowers throughout the summer.

Traditional/Medicinal Use Comfrey was long used to set broken bones. A boiled mash was made from the roots and applied to a broken finger, arm, or leg. Within a short period of time, the mash would set as strong as any plaster.

Edibility The plant's roughness and hairiness make it a little tough to eat raw. However, when the young leaves are cooked like spinach, they are very palatable, either by themselves or mixed with other greens.

COMMON BURDOCK

Arctium minus

ASTER FAMILY Asteraceae

Description Common burdock, a biennial, starts its first year of life as a basal rosette. In its second year, it sends up a robust branched stem over 1.8 m (6 ft.) in height with similar width. Its violet-coloured flowers are surrounded by hooked bracts that later form the burr, up to 2.5 cm (1 in.) across. The heart-shaped basal leaves can be as long as 45 cm (18 in.); the stem leaves are smaller. Great burdock (*A. lappa*) is almost identical to common burdock, except with larger flowers set in flat-topped clusters instead of along the stem.

Origin Native to Eurasia, it was first documented in New England in 1638.

Etymology The common name burdock is from the burr-like seed heads and from the leaves' resemblance to curled dock (*R. crispus*).

Habitat Can be extremely invasive in moist lowland pastures and areas with nutrient-rich soils.

Season Flowers from July to autumn.

Traditional/Medicinal Use Burdock has been used in blood medicines and as a cure for eczema.

Edibility The young stalks can be peeled (thoroughly) and eaten raw. The first-year taproots have long been cultivated as a food source.

COMMON CAMAS

Camassia quamash

LILY FAMILY Liliaceae

Description Common camas is an herbaceous perennial from bulb up to 70 cm (20 in.) high. It has six beautiful blue-purple petals up to 4 cm (1.5 in.) across. The long grass-like leaves are slightly shorter than the flowering stem. Great camas (*C. leichtlinii*) is very similar but taller, up to 1.2 m (4 ft.) The best way to distinguish the two species is by the petals of the great camas, which twist around the fruit as they wither.

Habitat Moist meadows in spring, dry meadows in summer, and on grassy slopes at low elevations.

Season Flowers from mid-April to May.

Edibility The bulbs can be roasted or boiled and were an important Indigenous food source; wars were fought over ownership of certain meadows.

CAUTION When out of flower, common camas is easily mistaken for death camas (*T. venenosus*), which is entirely poisonous.

▶ COMMON CHICKWEED MISCHIEVOUS JACK

Stellaria media

PINK FAMILY Caryophyllaceae

Description Common chickweed is a shade-tolerant, mat-forming annual that grows to 80 cm (32 in.) across. Its star-like flowers are about 0.6 cm (0.25 in.) across, have five deeply notched petals, and are borne in both the leaf axils and in terminal clusters. The leaves are about 2.5 cm (1 in.) long, opposite, and sharply egg shaped. The stems have a single line of hairs running up one side.

Origin Introduced from Eurasia.

Etymology The genus name *Stellaria* refers to the star-like flowers. The common name refers to the leaves and seeds being liked by most birds, especially chickens.

Habitat Garden lawns, golf courses, roadsides, and ditches. Its ability to withstand shade makes it very invasive.

Season Chickweed is in flower year-round, though it is most abundant in the cooler and wetter months of September to April.

Traditional/Medicinal Use Chickweed has been used for hundreds of years to treat inflammation, skin rashes, insect bites, and general itchiness.

Edibility The leaves can be added to salads or used as potherbs. The seeds can be dried, ground, and mixed with flour.

COMMON DANDELION

LION'S TOOTH / WET-A-BED / SWINE'S SNOUT

Taraxacum officinale

ASTER FAMILY Asteraceae

Description The common dandelion is probably the most recognized weed. It is a taprooted perennial that grows to 50 cm (20 in.) in height. Yellow ray flowers are up to 5 cm (2 in.) across and are borne singly on a hollow stem. The wavy green leaves grow in a basal rosette and can reach lengths of 30 cm (12 in.) The flowers close as the sun disappears and reopen at the approach of sunrise.

Origin Introduced from Eurasia.

Etymology The digested green leaves are said to make the kidneys relax in children; hence the common name wet-a-bed. The common name dandelion is the adulterated English version of the French *dent de lion*, meaning "tooth of the lion." This supposedly refers to the sharply toothed leaves.

Habitat Almost anywhere below the sub-alpine level. Prefers deep, moist soils for its taproot.

Season Flowers throughout spring and summer, with a heavy floral display from April to May.

Traditional/Medicinal Use It is officially recognized in pharmacology as a remedy for kidney, liver, skin, and digestive problems.

Edibility All parts are edible, for tea, wine, beer, and cooking.

COMMON HAWKWEED

Hieracium vulgatum

ASTER FAMILY Asteraceae

Syn. *H. lachenalii*

Description Common hawkweed is an upright herbaceous perennial about 80 cm (31 in.) in height. Its flower heads are 2–3 cm (1–1.5 in.) across, have yellow ray florets only and no disc, and are borne in terminal clusters of 5 to 20. The basal leaves are arranged as a rosette, ovate stalked, and toothed, and the stem leaves are unstalked. The Pacific Northwest has several native hawkweeds that can be confused with the introduced species. Two natives are Canada hawkweed (*H. canadense*) and narrow-leaved hawkweed (*H. umbellatum*). A specialized key is necessary for proper identification. All species of hawkweed have a milky juice in their stems and leaves, and their flowers comprise only ray florets.

Origin Common hawkweed is a native of Europe.

Etymology The species name *vulgatum* means "common."

Habitat Spreads very rapidly in fields, roadsides, gardens, and lawns.

Season Flowers from July to September.

Traditional/Medicinal Use For centuries, hawkweeds have been used medicinally for diseases of the lungs, asthma, and consumption. The juice from the plant mixed with wine was thought to help digestion and dispel wind.

Edibility Young leaves can be used as a salad garnish.

▶ COMMON HOP EUROPEAN HOP / HOP

Humulus lupulus

HEMP FAMILY Cannabaceae

Description Common hops are herbaceous perennial vines capable of growing over 9 m (30 ft.) annually. The vines are dioecious—they have male and female flowers on separate plants. Both types of flower arise from the leaf axils. The male flowers are in loose panicles 5–12 cm (2–5 in.) long, and the female flowers are greenish yellow and cone-like, with overlapping bracts. The leaves are up to 10 cm (4 in.) long, opposite, lobed, and heart shaped.

Origin Introduced from Europe.

Etymology The genus name *Humulus* is thought to be derived from humus, the rich soil the vine likes to grow in. The common name hop is from the ancient Anglo-Saxon word *hoppan*, "climbing."

Habitat Loves humus-rich moist soils.

Season The hops ripen from August to September.

Traditional/Medicinal Use Over the centuries, the common hop has been used to treat heart disease, nervous disorders, toothaches, earaches, and stomach and digestion problems.

Edibility The young shoots can be prepared like asparagus. The fruit of the vines has been used in breweries since the fourteenth century. Many a fine ale and beer have been made from them.

COMMON HORSETAIL

Equisetum arvense

HORSETAIL FAMILY Equisetaceae

Description Common horsetail is an herbaceous perennial up to 75 cm (30 in.) in height. It has two types of stems, fertile and sterile, both hollow except at the nodes. The fertile stems are unbranched, up to 30 cm (12 in.) in height, and lack chlorophyll; they bear spores in the terminal head. The green sterile stems grow up to 75 cm (30 in.) in height and have leaves whorled at the joints. Horsetails are all that is left of the prehistoric Equisetaceae family, some members of which grew to the size of trees.

Habitat Low, wet seepage areas, meadows, damp sandy soils, and gravel roads from low to high elevations.

Season Spores are produced from May to July.

Edibility The young fertile shoots of both common horsetail and giant horsetail (*E. telmateia*) can be eaten raw or boiled.

▶ COMMON KNOTWEED

PROSTRATE KNOTWEED / KNOTGRASS / WIREWEED / BIRDWEED

Polygonum aviculare

BUCKWHEAT FAMILY Polygonaceae

Syn. *P. arenastrum, P. littorale*

Description Common knotweed is a mat-forming taprooted annual 0.3–1 m (1–3.3 ft.) long. Its tiny flowers are greenish with white to red margins, and grow in clusters of one to five in the leaf axils. The bluish-green leaves are alternate and elliptical, with short stalks.

Origin Native to Eurasia; thought to have come over at the beginning of the 1800s.

Etymology The species name *aviculare* is from *avis*, "bird," in reference to the many small birds that feed on the seeds. This is also the source of the common name birdweed.

Habitat Mainly seen in sidewalk cracks and on roadsides. However, it is becoming a problem in lawns, gardens, and grain fields.

Season Flowers from May to October.

Traditional/Medicinal Use The leaves have astringent and diuretic properties.

Edibility The leaves can be steeped for a tea, eaten raw in salads, or used as a potherb. They are rich in zinc. The seeds can be dried, ground, and used in flour.

COMMON ORACHE

SPEARSCALE / SPREADING ORACHE / WILD ARRACH / SPEAR SALTBUSH

Atriplex patula

GOOSEFOOT FAMILY Chenopodiaceae

Description Common orache is a fleshy taprooted annual up to 90 cm (3 ft.) high. Its tiny flowers are greenish and borne both in the leaf axils and in dense terminal spikes. The arrow-shaped lower leaves are opposite, while the upper leaves are alternating. The surfaces of the young leaves often have a mealy white powder that disappears as the leaves mature. The seeds are borne in conspicuous bumpy to smooth triangular bracts.

Origin Common orache most likely originated in Eurasia.

Etymology The species name *patula* means "spreading," referring to the plant's sometimes reclining character. The common name orache is from *aurago*, "golden"; the seeds were mixed with wine and used to treat yellow jaundice.

Habitat Common orache is mainly found near the ocean, just above the high-tide line, where Canada geese forage on it.

Season Flowers from July to September.

Traditional/Medicinal Use Tinctures made from the seeds were used to treat headaches, pain, and early rheumatism.

Edibility The young leaves can be sliced and added to salads or cooked like spinach. The seeds can be dried, ground, and mixed with flour.

COMMON SOW THISTLE

HARE'S THISTLE / HARE'S LETTUCE

Sonchus oleraceus

ASTER FAMILY Asteraceae

Description Common sow thistle is a taprooted annual up to 1.2 m (4 ft.) in height, more commonly seen at 0.6 m (2 ft.). Its yellow flowers are up to 2.5 cm (1 in.) across, have ray florets only, and are borne in flat or round-topped clusters. The flowers are very similar to those of annual sow thistle. The leaves are 5–30 cm (2–12 in.) long, with margins and sharply cut lobes. The base passes through the stem.

Origin Introduced from Europe.

Etymology The species name *oleraceus* is given to cultivated plants, vegetables, and potherbs. As one of the common names suggests, it is a favourite with rabbits.

Habitat Waste places, forest openings, pastures, and gardens.

Season Flowers from June to September.

Traditional/Medicinal Use Leaves and stalks were thought to be good for nursing mothers, probably because of the association of the milky juice produced from the broken leaves and stalks. The milky juice of all sow thistles is said to be a healthy wash for the skin.

Edibility Young leaves taste like dandelion and can be cooked like spinach.

COMMON STORK'S BILL HERON'S BILL / RED-STEM FILAREE

Erodium cicutarium

GERANIUM FAMILY Geraniaceae

Description Common stork's bill is a reddish branched annual or winter biennial, 5–25 cm (2–10 in.) in height. Its pink to lavender flowers are up to 1 cm (0.5 in.) across, have five petals, and are borne in terminal umbrella-like clusters. The leaves are stiffly hairy and pinnately compounded so fine that the leaflets appear fern-like. True geranium leaves are palmately compounded.

Origin Originally from the Mediterranean and introduced from Europe, most likely in a contaminated alfalfa bale.

Etymology The shape of the mature seed pods gives the plant its common name, stork's bill. The genus name *Erodium* is from the Greek word *erodios*, meaning "heron." This plant's name should be common heron's bill. The species name *cicutarium* means "resembling *cicuta*," or water hemlock, because the leaves of both plants are similar.

Habitat Roadsides, lawns, forest edges. Prefers a sandy soil.

Season All summer until first frost.

Edibility The young leaves can be eaten in salads or used as a potherb.

COMMON TANSY

TANSY / BUTTONS / BITTER BUTTONS / PARSLEY FERN

Tanacetum vulgare

ASTER FAMILY Asteraceae

Description Common tansy is an aromatic herbaceous perennial up to 1.5 m (5 ft.) in height. Its button-like yellow flowers have disc florets only, which are grouped together to form attractive flat-topped clusters. The feather-like leaves are 5–25 cm (2–10 in.) long, alternate, and finely divided.

Origin Eurasian temperate.

Etymology The common name tansy is thought to mean "immortal." This could be for two reasons: first, that the flowers last so long, and second, that tansy was used to preserve and protect dead bodies from corruption.

Habitat Sandy soils, vacant fields, roadsides—not a fussy weed.

Season Blooms mid-summer into autumn.

Traditional/Medicinal Use Small doses of tansy are said to be anti-flatulent and rid the stomach of worms, while large doses are a violent irritant of the abdominal organs. Infusions of the dried flowers and seeds were used to relieve gout.

Edibility An old custom of European bishops was to make tansy cakes for Easter consumption. It was also used in puddings, omelettes, and gravy.

CAUTION Some people get skin irritations from just touching the plants.

COMMON VETCH

Vicia sativa

PEA FAMILY Fabaceae

Syn. *V. angustifolia, V. sativa* ssp. *nigra*

Description Common vetch is an herbaceous perennial vine that can climb or weakly support itself up to 90 cm (3 ft.) in height. Its sweet pea-like flowers are up to 2.5 cm (1 in.) long, light to dark purple, and usually borne in pairs originating from the leaf axils. The seed pods can grow to almost 7 cm (3 in.) long and contain four to twelve seeds. The alternating leaves are pinnately compounded into four to eight pairs of leaflets and terminated with a branching tendril. Each leaflet is slightly notched and tipped with a slender bristle.

Origin Introduced from Europe as a forage and cover crop and for use in hay-making and green manure.

Etymology Common vetch has been cultivated for centuries, hence the species name *sativa*, meaning "cultivated" or "sown." It is a close relative of the fava bean or broad bean (*V. faba*).

Habitat Mainly seen in fields or areas with grasses or plants on which it can climb.

Season Flowers from April to August.

Edibility The leaves, pods, and seeds can be eaten raw or used for potherbs.

COOLEY'S HEDGE-NETTLE

Stachys cooleyae

MINT FAMILY Lamiaceae

Description Cooley's hedge-nettle is an herbaceous perennial up to 1 m (3.3 ft.) in height. Its purple-red flowers are trumpet-like with a lower lip; they grow up to 4 cm (1.5 in.) long and are grouped in terminal clusters. The leaves are mint-like with toothed edges, opposite, finely hairy on both sides, and up to 15 cm (6 in.) long. The stems are square and finely hairy.

Etymology Cooley's hedge-nettle was first documented in 1891 by Grace Cooley, a professor from New Jersey who saw it near Nanaimo.

Habitat Moist open forests and streamsides at low elevations.

Season Flowers from June to mid-July.

Edibility The young stems can be chewed with the fibres being spit out. The roots (rhizomes) can be boiled to make a refreshing beverage.

▶ COW PARSNIP INDIAN CELERY

Heracleum lanatum

CARROT FAMILY Apiaceae

Description Cow parsnip is a tall, hollow-stemmed, herbaceous perennial 1–3 m (3.3–10 ft.) high. Its small white flowers are grouped in flat-topped, umbrella-like terminal clusters up to 25 cm (10 in.) across. It produces numerous small, egg-shaped seeds, 1 cm (0.5 in.) long, with a pleasant aroma. The large woolly compound leaves are divided into three leaflets, one terminal and two lateral, up to 30 cm (12 in.) across. Giant cow parsnip (*H. mantegazzianum*) is an introduced species that grows to 4 m (13 ft.) in height and can be seen in urban areas.

Etymology The genus name *Heracleum* is fitting for this plant of Herculean proportions.

Habitat Moist forests, meadows, marshes, and roadsides from low to high elevations.

Season Flowering starts at the end of June in lower areas and at the end of July at higher elevations.

Edibility The stems can be peeled and eaten raw. However, the discarded skin is considered poisonous.

CAUTION Both species can cause severe blistering and rashes when handled.

CREEPING BELLFLOWER PURPLE BELL / GARDEN HAREBELL

Campanula rapunculoides

HAREBELL FAMILY Campanulaceae

Description Creeping bellflower is a perennial up to 90 cm (3 ft.) in height. Its bell-shaped flowers are 3.5 cm (1.5 in.) long, blue to light purple, nodding, and borne in the upper leaf axils. The lower leaves are up to 7 cm (3 in.) long, alternate, and long stalked. The upper leaves are reduced in size, stalkless, and lanceolate to egg shaped.

Origin A garden escapee; native to Eurasia.

Etymology The genus name *Campanula* means "little bell." The species name *rapunculoides* means "resembling rapunculus" or "rampion-like," which is diminutive of turnip (*B. rapa*), referring to the thick roots.

Habitat Abandoned parking lots, fields, roadsides, and gardens.

Season Flowers from July to September.

Edibility The roots can be prepared like asparagus, and the leaves can be used in salads. Both are said to be of low quality.

CREEPING JENNY

MONEYWORT / WANDERING JENNY / CREEPING JOAN / LOOSESTRIFE

Lysimachia nummularia

PRIMULA FAMILY Primulaceae

Description Creeping jenny is a prostrate herbaceous perennial capable of covering hundreds of square metres or feet by rooting at the nodes and its spreading rhizomes. The cup-shaped flowers have five deeply-lobed yellow petals, which are borne singly on short stalks arising from the leaf axils. The leaves are up to 2.5 cm (1 in.) long, opposite, and oblong to circular in outline.

Origin Introduced from Europe as an ornamental groundcover.

Etymology The species name *nummularia* means "coin shaped," referring to the shape of the leaves, which is also the source of its common name, moneywort. Long ago, creeping jenny was called *serpentaria* from the belief that wounded snakes would lie on the leaves to heal their wounds.

Habitat Moist soils, mainly in cultivated gardens and containers.

Season Flowers from June to July.

Traditional/Medicinal Use The leaves and flowers can be steeped for an herbal tea. The bruised leaves were at one time applied to cuts and wounds for a blood stanch.

Edibility The leaves and flowers can be steeped for an herbal tea.

CURLED DOCK NARROW DOCK / SOUR DOCK

Rumex crispus

BUCKWHEAT FAMILY Polygonaceae

Syn. *Rumex elongates*

Description Curled dock is a taprooted herbaceous perennial up to 1 m (3.3 ft.) in height. Its small, unattractive flowers are greenish and borne in long terminal clusters. The spent flowers and seeds persist on the stalks through autumn and early winter. The basal leaves are up to 30 cm (12 in.) long, with curly edges, and the upper leaves are reduced in size.

Origin Introduced from Eurasia.

Etymology The long narrow leaves with curled margins make this dock one of the easiest to identify. The species name *crispus* means "curly."

Habitat Can be extremely invasive in moist lowland pastures and areas with nutrient-rich soils.

Season Flowers from May through the summer.

Traditional/Medicinal Use The taproots of curled dock have been used medicinally for centuries. Its astringent properties were thought to make it a good blood cleanser.

Edibility The young leaves can be used in salads or as potherbs. The seeds can be dried, ground, and mixed with flour.

▶ CUT-LEAVED GERANIUM

CUT-LEAVED CRANE'S BILL

Geranium dissectum

GERANIUM FAMILY Geraniaceae

Description Cut-leaved geranium is a taprooted annual 15–60 cm (6–24 in.) in height. Its flowers are up to 0.5 cm (0.3 in.) across, pink to purple, and usually borne in pairs on long stalks. The fruit capsules are up to 2 cm (1 in.) long, five parted, and hairy. The leaves are up to 5 cm (2 in.) across, opposite, and palmately dissected into five segments.

Origin Introduced from Eurasia.

Etymology The genus name *Geranium* means "crane," an allusion to the pointed beaks of the fruit capsules. A soft-brown dye can be obtained from steeping the dry flowers.

Habitat Fields, abandoned lawn areas, and forest edges.

Season Flowers from May to September.

Traditional/Medicinal Use An infusion made from the leaves and roots was used to treat diarrhea in children.

Edibility The roots were cooked and used as famine food.

DAME'S ROCKET

DAME'S VIOLET / SWEET ROCKET / PURPLE ROCKET

Hesperis matronalis

MUSTARD FAMILY Brassicaceae

Description Dame's rocket is a taprooted biennial or short-lived herbaceous perennial 0.6–1.2 m (2–4 ft.) in height. Its velvet-like flowers may be white, pink, purple, or variegated. The lanceolate leaves can be as long as 20 cm (8 in.) However, they are clasping and reduced in size upward.

Origin A garden escapee from Eurasia.

Etymology The genus name *Hesperis* means "evening," in reference to the flowers becoming far more fragrant in the evening. Long ago, the flowers were a favourite of matrons, hence the species name *matronalis*. From this expression later came the corrupted dame's violet. In medieval times, the word violet meant "flower."

Habitat Outside of cultivated gardens, it prefers areas where the lawn mower can't get it, forest edges, and the perimeters of golf courses.

Season Flowers from the end of March to summer.

Edibility The young leaves, though bitter, can be used in salads.

DWARF MALLOW

COMMON MALLOW / CHEESES / CHEESE WEED

Malva neglecta

MALLOW FAMILY Malvaceae

Description Dwarf mallow is a short-lived herbaceous perennial up to 60 cm (24 in.) tall. Its flowers are up to 2.5 cm (1 in.) across, white to pale purple, and borne on short stalks originating from the axils. The long-stalked leaves are up to 5 cm (2 in.) across, five to seven lobed, and rounded to kidney shaped.

Origin Introduced from Eurasia.

Etymology The seeds are borne in round, flat carpels that somewhat resemble a wheel of cheese, hence the common names cheese weed and cheeses. The species name *neglecta* means "overlooked," in reference to the plant's many uses and benefits being overlooked.

Habitat Roadsides, ditches, abandoned lawns, laneways. Easily identified when in flower.

Season Flowers from April to October.

Edibility The young leaves and seeds can be eaten raw, as potherbs. A tea can also be made from the dried leaves, flowers, or roots. See musk mallow for more uses.

EARLY WINTER CRESS

SCURVY GRASS / SCURVY CRESS / BARBARA'S HERB / BELLE ISLE CRESS

Barbarea verna

MUSTARD FAMILY Brassicaceae

Syn. *B. praecox*

Description Early winter cress is a taprooted biennial 30–90 cm (1–3 ft.) in height. Its bright-yellow flowers are borne on a terminal flower stalk and flower stalks originating in the leaf axils. The pinnately lobed basal leaves are long stalked, while the upper leaves are reduced in size and clasping.

Similar Species American winter cress (*B. orthoceras*) is very similar in appearance, but its seed pods (siliques) are shorter: 2.5–5 cm (1–2 in.) long, compared with 5–8 cm (2–3 in.) long on early winter cress.

Origin Introduced from Eurasia.

Etymology The common name cress is Old English for "to eat" or "to nibble on." The genus name *Barbarea* commemorates the fourth-century saint Barbara (d. 306), probably because early winter cress was eaten at her winter festivals.

Habitat Beside farmland and abandoned fields are the best places to find this early producer.

Season Flowers from April to June.

Traditional/Medicinal Use Early winter cress is high in both vitamin A and vitamin C and was used as an antiscorbutic—a food or medicine that helps prevents scurvy, thus the names scurvy grass and scurvy cress.

Edibility It has been cultivated for centuries as a vegetable and is usually cooked like spinach. There are sources still selling the seeds.

EDIBLE THISTLE

Cirsium edule

ASTER FAMILY Asteraceae

Description Edible thistle is a showy biennial and sometimes perennial growing as tall as 2 m (6.6 ft.) in favourable conditions. The well-armed leaves are alternating and lance shaped, with spined lobes. The beautiful pinkish-purple flowers nod when young. Short-style thistle (*C. brevistylum*) is similar, except the styles are shorter.

Habitat Lightly covered forest edges and moist meadows.

Season Flowers from July to September, depending on elevation.

Edibility As with most thistles native to the Pacific Northwest, the taproots and lower stem provide emergency food when peeled.

ENGLISH DAISY COMMON DAISY

Bellis perennis

ASTER FAMILY Asteraceae

Description English daisy is a low-growing herbaceous perennial with flower stalks reaching up to 15 cm (6 in.) in height. Its very familiar white and yellow flowers are approximately 2.5 cm (1 in.) across when they are open; they shut at night and in damp weather. The long-stalked leaves are formed in a basal rosette and range in shape from oval to spoon shaped.

Origin Introduced from Europe.

Etymology The Latin name *Bellis perennis* translates to "pretty perennial," an apt name for this beautiful little plant. The common name daisy is a corruption of the Old English name "day's eye," so called because it is open during the day and shut at night.

Habitat Usually a welcome guest in private lawns, though lawn-keepers at golf courses may have a different opinion. Likes nutrient-rich moist soils.

Season English daisy has a prolific flower display in late spring and a modest showing through autumn.

Traditional/Medicinal Use The plant was very popular in the fourteenth century for its use as a mild laxative and antispasmodic, and for stomach and intestinal problems.

Edibility The flowers were used to make a summer wine, and the leaves were used as salad greens and potherbs.

EVENING PRIMROSE

PRIMROSE / EVENING STAR / FEVER PLANT / KING'S CURE-ALL

Oenothera biennis

EVENING PRIMROSE FAMILY Onagraceae

Description Evening primrose is a night-scented biennial up to 1.2 m (4 ft.) in height. Its bright yellow flowers are up to 5 cm (2 in.) across, have four petals, and are borne in leafy terminal spikes. The stem leaves are up to 10 cm (4 in.) long, alternate, lance shaped, and reduced in size upward. The seeds are contained in upright, hairy capsules up to 3 cm (1 in.) long.

Origin Native to the east coast of North America.

Etymology The common name evening primrose refers to the flowers fully opening at night; their sweet scent attracts nighttime pollinating moths. Once cultivated for its edible roots, it is now grown primarily for the oil contained in the seeds.

Habitat Commonly seen as a roadside attraction, in ditches and abandoned fields. Prefers arid, well-drained soils.

Season Flowers from June to September.

Traditional/Medicinal Use For centuries, evening primrose was known as king's cure-all. It has been used to treat coughs, depression, cyclic nostalgia, high blood pressure, schizophrenia, eczema, diabetes, and premenstrual syndrome. Evening primrose oil is used in cosmetic creams, soaps, lip balms, lotions, and aromatherapy and massage carrier oils.

Edibility The leaves, flowers, seeds, and first-year roots can be eaten raw or as potherbs. The root is most often sought after and is cooked like carrots.

▶ EVERGREEN VIOLET

Viola sempervirens

VIOLET FAMILY Violaceae

Description Evergreen violet is a small creeping perennial up to 8 cm (3 in.) in height, the smallest of the yellow flowering violets in the Pacific Northwest. Its yellow flowers are solitary and 1–5 cm (0.5–2 in.) across, with delicate brown veins on the bottom petals. The evergreen leaves grow to 3 cm (1 in.) across and are broadly heart shaped and leathery. It spreads by sending out slender trailing stems.

Habitat Dry to moist forests at low to mid elevations. Common in forested areas, often hiding under bushes or fallen leaves.

Season Flowering starts mid-March.

Traditional/Medicinal Use It is said that violets worn around the head will dispel the fumes of wine and prevent headache and dizziness.

Edibility The flowers and leaves from all true violets are edible. The colourful flowers make a good salad garnish.

FAIRYSLIPPER

Calypso bulbosa

ORCHID FAMILY Orchidaceae

Description Fairyslipper is a delicate herbaceous perennial from corms up to 20 cm (8 in.) in height. Its flower is light purple; the lower lip is lighter and decorated with spots, stripes, and coloured hairs. The single leaf is broadly lanceolate and withers with the flower; a new leaf appears in late summer and remains through the winter. This is one of the most beautiful of the Pacific Northwest's native orchids.

Habitat Mostly associated with Douglas fir and grand fir forests.

Season Flowers from April to May.

Edibility When boiled, the corms have a rich, buttery flavour; Haida people ate fairyslipper corms in small quantities.

FALSE LILY OF THE VALLEY

Maianthemum dilatatum

LILY FAMILY Liliaceae

Description False lily of the valley is a small herbaceous perennial up to 30 cm (12 in.) tall. Its small white flowers appear in April to May, clustered on 5–10 cm (2–4 in.) spikes. The slightly fragrant flowers are quickly replaced by berries 0.6 cm (0.25 in.) across; the berries go through summer a speckled green and brown but turn ruby red by the autumn. The dark-green leaves are alternate, heart shaped, and slightly twisted, up to 10 cm (4 in.) long.

Etymology The genus name *Maianthemum* is from the Greek words *Maios*, "May," and *anthemion*, "blossom."

Habitat Moist coastal forests at low elevations.

Season Flowers bloom in early May, and berries start showing mid-June.

Traditional/Medicinal Use Haida people put the leaves on cuts and wounds, while Cowichan people drank root infusions for internal injuries. Leaves and roots were made into poultices.

Edibility The berries were eaten by several groups of coastal Pacific Northwest Indigenous people. The berries were not considered a great gastronomic experience. They were probably consumed when food rations were low.

FENNEL WILD FENNEL

Foeniculum vulgare

CARROT FAMILY Apiaceae

Description Fennel is an herbaceous perennial up to 1.8 m (6 ft.) in height. Its beautiful golden flowers are borne in large, flat, terminal umbels up to 10 cm (4 in.) across. The seeds are ribbed, elliptical, and up to 0.5 cm (0.2 in.) long, which is half the length of the cultivated species, *F. officinalis*. The bright-green leaves are dissected so finely they give the plant a feathery appearance.

Origin Native to Europe but considered more precisely indigenous to the shores of the Mediterranean Sea. Ancient Italians are thought to be responsible for its spread. Wherever they colonized, fennel was to be found. Its use can be traced back to Roman times.

Etymology The genus name *Foeniculum* is Latin for "fennel," and the species name *vulgare* means "common."

Habitat Considered a garden escapee, it likes well-drained soils. It can be seen in back alleys, cracks in sidewalks, and areas surrounding farmland.

Season Flowers from July to September.

Traditional/Medicinal Use Tea made from the seeds was used as a carminative.

Edibility The seeds, leaves, and roots have an anise flavour and are edible. The seeds were generally used to flavour soups, stews, and tea. The leaves and stems have long been cooked or boiled with fish.

FIELD PEPPERGRASS FIELD PEPPERWORT / COW CRESS

Lepidium campestre

MUSTARD FAMILY Brassicaceae

Syn. *Neolepia campestre, Thlaspi campestre*

Description Field peppergrass is a short, densely hairy annual or biennial up to 50 cm (20 in.) tall. Its tiny white flowers have yellow stamens and four petals and are borne in long terminal clusters. The seed pods (siliques) are up to 0.6 cm (0.25 in.) long and egg shaped, with a notch at the top. The basal leaves are stalked and sit in a rosette. The stem leaves are unstalked, reduced in size, and clasping. The seed pods of field pennycress (*Thlaspi arvense*) are similar to those of field peppergrass, except they are two to three times larger.

Origin Introduced from Eurasia.

Etymology The species name *campestre* means "from the pasture."

Habitat Farmland, abandoned fields. Prefers rich soils.

Season Flowers from May to July.

Edibility The seeds have a peppery taste. They make a good mustard paste if ground and mixed with vinegar or white wine. The young shoots were used as a substitute for watercress (*N. officinale*).

FIREWEED

Epilobium angustifolium

EVENING PRIMROSE FAMILY Onagraceae

Description Fireweed is a tall herbaceous perennial that reaches heights of 3 m (10 ft.) in good soil. Its purple-red flowers grow on long showy terminal clusters. The leaves are alternate, lance shaped like a willow's, 10–20 cm (4–8 in.) long, and darker green above than below. The minute seeds are produced in pods 5–10 cm (2–4 in.) long and have silky hairs for easy wind dispersal. Fireweed flowers have long been a beekeeper's favourite.

Etymology The common name fireweed comes from the fact that it is one of the first plants to grow on burned sites; it typically follows wildfires.

Habitat Common throughout BC in open areas and at burned sites.

Season Flowers from June to July at high elevations.

Traditional Use The stem fibres were twisted into twine and made into fishing nets, and the fluffy seeds were used in padding and weaving.

Edibility The young stems and shoots can be consumed as a green vegetable and the flowers used to brighten up salads.

FOOL'S ONION

Brodiaea hyacinthina

LILY FAMILY Liliaceae

Description Fool's onion is an herbaceous perennial from corms up to 30 cm (12 in.) in height. The small star-shaped flowers are white with fine stripes of green and held up in terminal clusters on thin stems up to 60 cm (24 in.) long. The grass-like leaves are basal and have usually disappeared by the time the flowers are noticeable.

Habitat Rocky outcrops and grassy slopes.

Season Flowers at the beginning of June.

Edibility The corms can be eaten raw or boiled.

GARLIC MUSTARD HEDGE GARLIC / JACK-IN-THE-HEDGE

Alliaria petiolata

MUSTARD FAMILY Brassicaceae

Description Garlic mustard is a taprooted biennial 0.3–1.2 m (1–4 ft.) in height. Its white flowers have four petals and are borne in terminal clusters. The seed pods (siliques) that are produced from these are up to 5 cm (2 in.) long and upright, and contain tiny blackish seeds. The basal leaves are on long stalks (petioles) and are kidney shaped, while the stem leaves are heart shaped and shorter stalked.

Origin Introduced from Eurasia by the early settlers; one of the first records of it is from 1868 on Long Island, New York.

Etymology The genus name *Alliaria* means "garlic smelling," which it is.

Habitat Likes moist rich soils at forest edges, fields, and abandoned farmland.

Season Flowers from April to June.

Traditional/Medicinal Use The juice from the plants was used externally to treat wounds and ulcers. The leaves were chewed for sore gums and gum ulcers.

Edibility The leaves, flowers, seeds, and roots are edible, either raw or cooked. The garlic odour and taste from the plant were used to spice up soups, stews, roasts, and fish.

GIANT HOGWEED

Heracleum mantegazzianum

CARROT FAMILY Apiaceae

Description Giant hogweed is a robust herbaceous perennial up to 4.5 m (15 ft.) tall. Its small white flower clusters fit together much like an umbrella, forming a giant inflorescence 0.6–0.9 m (2–3 ft.) across. The leaves are coarsely toothed and similar to a maple leaf, except that they attain sizes of 0.6–1.5 m (2–5 ft.) across. Our native cow parsnip (*H. lanatum*) is similar but smaller, 1–3 m (3.3–10 ft.) in height.

Origin Giant hogweed was brought over from the Caucasus Mountains of Asia as a garden curiosity. The earliest records of its being on the West Coast are from 1950.

Etymology Hogweed's ability to grow so strongly has given it its genus name *Heracleum* in honour of Hercules. The species name is in honour of Paolo Mantegazza (1831–1910), an Italian anthropologist and traveller. Its common name derives from the fact that the English fed the leaves to hogs.

Habitat Moist soils in meadows, large parks, ditches, and forest edges.

Season Flowers from June to July.

Edibility Iranians used the seed as a spice (golpar). Unless you are very hungry or very curious, I would not recommend eating this plant. If you do want to try this plant, please wear rubber gloves when collecting (see caution below). When peeled, the stalks and stems are just like our domesticated celery. The stalks and stems should be harvested before the flowers open.

CAUTION The plant can cause serious skin rashes when in contact with the skin and subsequently exposed to sunlight (phototoxicity). I have had rashes on my hands and arms from collecting stems while not wearing gloves or long sleeves.

GIANT KNOTWEED

SPREADING KNOTWEED / KNOTWEED

Polygonum sachalinense

BUCKWHEAT FAMILY Polygonaceae

Syn. *Reynoutria sachalinense, Fallopia sachalinense*

Description Giant knotweed is a rhizomatous herbaceous perennial 1.8–3.6 m (6–12 ft.) in height. Its greenish-white flowers are up to 0.3 cm (0.1 in.) across and borne in dense panicles arising from the leaf axils. The stems are reddish brown and hollow, with bamboo-like joints. The leaves are 10–30 cm (4–12 in.) long, alternating, and roundly egg shaped.

Similar Species Japanese knotweed (*P. cuspidatum*) is very similar in appearance and aggressiveness. It is generally seen as a smaller plant with leaves that are abruptly pointed.

Origin Japan.

Etymology The species name *sachalinense* refers to Sakhalin Island, north of Japan, where giant knotweed is native.

Habitat Likes deep moist soils at forest edges, parks, lawns, and abandoned lots.

Season Flowers from late June to October.

Medicinal Use: Resveratrol is an active component extracted from Japanese knotweed and red grapes that is being looked at to treat AIDS and cancer.

Edibility Both species of giant knotweed are excellent vegetables. The new shoots with unfurled leaves can be treated as asparagus, and the rhizomes can be boiled or roasted. Be careful not to overboil the new shoots and leaves as they will turn to mush very quickly.

GOAT'S BEARD

Aruncus dioicus

ROSE FAMILY Rosaceae

Description Goat's beard is a deciduous shrub 3 m (10 ft.) in height. The plants are dioecious: male and female flowers appear on separate plants. The tiny white flowers are compacted into hanging panicles up to 60 cm (24 in.) long. The leaves are compounded three times (thrice pinnate), and the leaflets are bright green with a toothed edge, tapering to a point. With a little imagination, the hanging flowers look like a goat's beard.

Habitat Moist open woodlands, creeksides, and wet rocky slopes at lower elevations.

Season Flowers from June to early July.

Traditional/Medicinal Use The roots were steeped, and the warm tea given to an expecting woman just before giving birth. It was thought the tea would help her heal.

Edibility Young shoots and leaves can be boiled and eaten; however, it cannot be safely eaten raw.

GROUND-IVY

CREEPING CHARLIE / ALEHOOF / GILL-OVER-THE-GROUND

Glechoma hederacea

MINT FAMILY Lamiaceae

Syn. *Nepeta glechoma*

Description Ground-ivy is a semi-evergreen trailing perennial, 40 cm (16 in.) long. Its bluish-purple flowers are up to 2.5 cm (1 in.) long and trumpet shaped with two lips. The flowers are borne three to four in whorls that originate in the upper leaves. The leaves are 2.5–5 cm (1–2 in.) long, kidney shaped, and opposite, with rounded teeth. The square stems root at the nodes when trailing on the ground. Ground-ivy will lose its leaves in areas with hard frost.

Origin Ground-ivy is thought to be one of the first edible herbal plants brought over from Europe by settlers in the 1600s.

Etymology As its common name alehoof suggests, ground-ivy was used to flavour and clarify beer. Gill-over-the-ground comes from the French word *guiller*, "to ferment beer."

Habitat Shaded woodland areas with rich moist soil.

Season Flowers from March to July.

Traditional/Medicinal Use Medicinally, ground-ivy can be dated back to the first century AD—it was known as a cure-all (panacea). Infusions of the leaves have been used to treat diseases of the lungs and kidneys, asthma, and jaundice, and to reduce fever and chronic coughs.

Edibility The leaves can be eaten raw and mixed with other salad greens.

▶ HAIRY BITTERCRESS

Cardamine hirsuta

MUSTARD FAMILY Brassicaceae

Description Hairy bittercress is a thinly taprooted annual 5–30 cm (2–12 in.) in height. The tiny white flowers are borne in compact racemes that thin out along the stem as the seeds are produced. The upright seed pods (siliques) are up to 2.5 cm (1 in.) long. The basal rosette of leaves is pinnately divided in two or more pairs of opposite-facing leaflets. The few stem leaves are unstalked and reduced in size.

Origin Introduced from Eurasia.

Etymology The genus name *Cardamine* was the name Greek botanist Pedanius Dioscorides (40–90 AD) used for cress, and the common name cress means "to eat" or "to nibble on." The species name *hirsuta* means "hairy."

Habitat Prefers cultivated soils but can be seen by train tracks, abandoned lots, vegetable gardens, and roadsides.

Season Flowers in spring and autumn, depending on when the seeds germinate.

Edibility As the common name suggests, the leaves are bitter. However, they were nibbled on and used in soups.

HAIRY CAT'S EAR FALSE DANDELION

Hypochaeris radicata

ASTER FAMILY Asteraceae

Description Hairy cat's ear is a taprooted herbaceous perennial 15–60 cm (6–24 in.) tall. Its yellow flowers are up to 4 cm (1.5 in.) across, have ray florets only (no disc), and are borne in terminal clusters on leafless stalks (as opposed to the dandelion's flowers, which are borne singly). The leaves are 5–20 cm (2–8 in.) long, obovate, round toothed, round lobed, densely hairy, and formed in a basal rosette.

Origin Native to Europe.

Etymology With considerable imagination, one can see a resemblance between the hairy leaves and a cat's ear.

Habitat Does very well in lawn areas in parks, fields, and waste areas. Does not require irrigation to compete with native flora.

Season Flowers from May to October, with the main floral display in August.

Edibility The young leaves can be eaten raw or cooked. I find the young leaves slightly bitter, even after cooking.

HARVEST LILY

Brodiaea coronaria

LILY FAMILY Liliaceae

Description Harvest lily is an herbaceous perennial from corms, growing to 30 cm (12 in.) in height. Its purple, trumpet-shaped flowers are 4 cm (1.5 in.) long and grow in clusters of three to five. The leaves are grass-like and wither by the time flowers are noticeable.

Habitat Prefers well-drained grassy slopes.

Season Flowers from the end of June to July.

Edibility The corms can be boiled or steamed and were often harvested for winter consumption.

HEDGE MUSTARD SINGER'S PLANT / CRAMBLING ROCKET

Sisymbrium officinale

MUSTARD FAMILY Brassicaceae

Description Hedge mustard is a taprooted annual up to 90 cm (3 ft.) tall. Its pale-yellow flowers have four petals and are borne in clusters that elongate as the seed is produced. The seed pods (siliques) are up to 1.5 cm (0.75 in.) long, erect, and held tightly to the stem. The basal leaves are up to 20 cm (8 in.) long, pinnately divided, hairy, and reduced in size upward.

Origin Native to Eurasia.

Habitat Can grow in almost any soil, from roadsides and waste areas to wonderfully cultivated gardens.

Season Flowers from April to July.

Traditional/Medicinal Use Infusions of the plant were used to treat throat illnesses. The French called it the singer's plant. It was used to cure the loss of voice, as happened to many singers.

Edibility The young leaves, picked before the flowers, are good in salads and cooked as potherbs.

HOOKER'S ONION

Allium acuminatum

LILY FAMILY Liliaceae

Description Hooker's onion is an herbaceous perennial up to 30 cm (12 in.) in height. Its rose-coloured flowers are held in upright umbels, unlike those of nodding onion (*A. cernuum*). The leaves are grass-like and wither by blooming time. When crushed, the entire plant smells like onion.

Etymology The species name *acuminatum* refers to the tapering flower petals.

Habitat Dry grassy slopes and crevices at low elevations.

Season Flowers from mid-May to June.

Edibility The small bulbs can be eaten raw or steamed.

IVY-LEAVED TOADFLAX

Cymbalaria muralis

FIGWORT FAMILY Scrophulariaceae

Syn. *Linaria cymbalaria*

Description Ivy-leaved toadflax is a delicate trailing or climbing vine up to 90 cm (36 in.) in length. Its violet-blue flowers are 1 cm (0.5 in.) long, yellow throated with two lobes, and borne singly on long stalks. The egg-shaped seeds are borne in rounded capsules. The leaves are up to 2.5 cm (1 in.) across, long stalked, and palmately rounded, with five to seven lobes.

Origin A garden escapee from Eurasia.

Etymology The genus name *Cymbalaria* is from the Greek word *kymbalon*, meaning "cymbal," referring to the shape of the leaves. The species name *muralis* means "walls" and describes where the plant commonly grows.

Habitat Grows in crevices in between stones and bricks in patios and walls. I have photographed this beautiful little plant on Hadrian's Wall, the Great Wall of China, and castle walls across Europe.

Season Flowers from April to November.

Traditional/Medicinal Use Poultices made from the leaves have been used as a blood stanch.

Edibility The leaves are bitter; however, they can be mixed with other greens in salads.

KINNIKINNICK BEARBERRY

Arctostaphylos uva-ursi

HEATHER FAMILY Ericaceae

Description Kinnikinnick is a trailing, mat-forming evergreen that rarely grows above 25 cm (10 in.) in height. Its fragrant pinkish flowers bloom in spring and are replaced by bright-red berries 1 cm (0.5 in.) across by late summer. The small oval leaves grow up to 3 cm (1 in.) long and are leathery and alternate. Grouse and bears feed on the berries.

Etymology Kinnikinnick is an eastern Indigenous word used to describe a tobacco mix.

Habitat Dry rock outcrops and well-drained forest areas from sea level to high elevations.

Season Flowers and berries from June to September, depending on elevation.

Traditional Use The leaves were dried and smoked in pipes made from hollowed branches. The dried leaves were sometimes mixed with other plants to extend their smoking supply.

Edibility The berries are a wee bit dry; however, they can be fried or roasted in butter to offset that.

LADY'S BEDSTRAW YELLOW BEDSTRAW / CHEESE RENNET

Galium verum

MADDER FAMILY Rubiaceae

Description Lady's bedstraw is a sprawling herbaceous perennial 90 cm (36 in.) in length. Its tiny yellow flowers are 0.3 cm (0.1 in.) across and borne in dense terminal panicles on square wiry stems. The thread-like leaves are 1–5 cm (0.5–2 in.) long and borne in whorls of four to eight.

Origin Introduced from Eurasia.

Etymology The genus name *Galium* is from the Greek word *gala*, meaning "milk."

Habitat Mainly seen as a weed in cultivated fields.

Season Flowers from July to August.

Traditional/Medicinal Use The dried stems, leaves, and flowers are antispasmodic and diuretic. It has been strewn on flowers and furniture for a flea deterrent, used as stuffing for mattresses, made into dye, and used for curdling milk.

Edibility The leaves and seeds can be eaten raw or as a potherb. A version of coffee can be made from roasting the seeds.

LADY'S THUMB SPOTTED KNOTWEED / COMMON SMARTWEED

Polygonum persicaria

BUCKWHEAT FAMILY Polygonaceae

Syn. *Persicaria maculata, P. maculosa, P. vulgaris*

Description Lady's thumb is a taprooted annual 30–90 cm (12–36 in.) in height. Its small pink flowers are borne in congested terminal racemes. The seeds are brown black and three sided. The leaves are up to 10 cm (4 in.) long, alternating, and lance shaped, and usually have a dark splotch (hence the common name lady's thumb) in the centre.

Origin Introduced from Eurasia.

Etymology The genus name *Polygonum* is from the Greek *polys*, "many," and *gony*, "knee," referring to the jointed stems. The species name *persicaria* means "from Persia," which is where botanists believe it originated.

Habitat Very common around ponds and irrigated fields and gardens.

Season Flowers from June to July.

Traditional/Medicinal Use The leaves are astringent and diuretic.

Edibility The peppery-tasting leaves and seeds can be eaten raw or as a potherb.

LAMB'S QUARTER

PIGWEED / WHITE GOOSE-FOOT / WILD SPINACH

Chenopodium album

GOOSEFOOT FAMILY Chenopodiaceae

Syn. *C. lanceolatum, C. dacoticum*

Description Lamb's quarter is a taprooted annual 60–90 cm (2–3 ft.) in height. Its tiny flowers are greenish grey and borne in dense clusters along the branch tips and leaf axils. The leaves are up to 10 cm (4 in.) long, fleshy, and shaped like lobed triangles or diamonds.

Origin Introduced from Europe. In 1965, the seeds of lamb's quarters were discovered in a Danish bog and dated back to 200 AD. There has been no evolutionary change in almost 2,000 years.

Etymology The genus name *Chenopodium* is Greek for "goose foot," though the leaf does not look much like a goose's foot. The species name *album* refers to the white mealy or powdery scales on the leaves and stems. The common name lamb's quarter is from the plant's flowering on August 1, the first day of the Lammas Day festival.

Habitat Seen in waste areas, vacant lots, and large parks.

Season Flowers from May to October.

Edibility The young leaves can be used in salads or cooked as spinach.

CAUTION It is not advisable to eat the leaves in large quantities or over long periods of time, as they contain oxalate salts.

LITTLE HOP CLOVER

SUCKLING CLOVER / SHAMROCK / YELLOW CLOVER / LESSER YELLOW TREFOIL

Trifolium dubium

PEA FAMILY Fabaceae

Syn. *T. minus, T. filiforme*

Description Little hop clover is a mat-forming annual 15–50 cm (6–20 in.) long. Its rounded flower heads have up to 20 lemon-yellow florets. As the flower heads mature, they turn brown and produce small pods bearing one seed. The leaves are up to 1 cm (0.5 in.) long, finely toothed on the upper half, and trifoliate (they have three leaflets).

Origin Introduced from Europe as a forage crop and soil conditioner.

Etymology Little hop clover is often considered the original shamrock used by St. Patrick; hence the common name shamrock. The species name *dubium* means "doubtful," referring to the uncertainty of its being an independent species. The common name little hop refers to the flower heads' resemblance to the cones of the hop plant.

Habitat Open areas with sunshine, street boulevards, playing fields, and gardens.

Season The seeds germinate from January to February and flower as early as April. The flowers seem to disappear in the summer heat and reappear with the autumn rains.

Edibility The young, tender tips can be eaten by themselves or added to a salad.

MARSHPEPPER SMARTWEED

Polygonum hydropiper

BUCKWHEAT FAMILY Polygonaceae

Description Marshpepper smartweed is a fibrous-rooted annual 30–90 cm (12–30 in.) in height. Its flowers are greenish pink and borne in long slender racemes that sometimes droop at the tip. The leaves are 8 cm (3 in.) long, alternating, lance shaped, and fringed with hairs.

Origin Introduced from Eurasia.

Etymology The species name *hydropiper* literally means "waterpepper," referring to it growing in water and tasting like pepper. It is usually seen growing in areas that are flooded from autumn to spring. The common name smartweed refers to the smarting sensation a person receives when handling the plants and then rubbing their eyes or from eating the leaves.

Habitat The muddy bottoms and banks of seasonal ponds and marshes.

Season Flowers from July to August.

Traditional/Medicinal Use Infusions and decoctions made from the leaves have been used as a stimulant, diuretic, antiseptic, and desiccant.

Edibility The young leaves and shoots have a peppery taste and can be eaten raw or as a potherb. Mature leaves can also be eaten, but have a much stronger flavour and will "smart" the tongue more as a result.

MOUNTAIN SORREL

Oxyria digyna

BUCKWHEAT FAMILY Polygonaceae

Description Mountain sorrel is a rather interesting plant you will likely come across when hiking the coastal mountains. Its flowers are greenish to pinkish and formed in dense panicles up to 20 cm (8 in.) long. The green leaves are rounded, up to 4 cm (1.5 in.) across, and held out on long arching stems 10–50 cm (4–20 in.) long. The leaves are sour tasting but are edible in small amounts.

Habitat Open rocky sites at mid to high elevations.

Season Flowers from July to September.

Edibility The leaves can be cooked as a potherb or mixed with honey and water for a refreshing drink. However, the leaves contain oxalates and should be consumed with moderation when eating raw.

MOUSE-EARED CHICKWEED

Cerastium fontanum

PINK FAMILY Caryophyllaceae

Syn. *Cerastium vulgatum*

Description Mouse-eared chickweed is a sprawling herbaceous biennial or short-lived perennial 50 cm (18–19 in.) in length. Its tiny white flowers have five deeply notched petals and are borne in terminal clusters. The leaves are 2.5 cm (1 in.) long, opposite, mouse-ear shaped, and coarsely hairy, like the stems.

Origin Introduced from Eurasia.

Etymology The genus name *Cerastium* is from the Greek word *kerastes*, "horned," referring to the tiny horned seed capsules. The species name *fontanum* means "growing by springs" or "growing by water."

Season Flowers from April to November.

Habitat Commonly seen in irrigated playing fields, lawns, and forest edges.

Edibility The young leaves and stems can be added to a salad or used as a potherb.

MUGWORT

Artemisia vulgaris

ASTER FAMILY Asteraceae

Description Mugwort is a very aromatic herbaceous perennial up to 1.5 m (5 ft.) in height. Its tiny flowers are bell shaped, woolly, reddish brown, and massed in upright panicles. The leaves are variable in shape, most being dissected to the main rib, green above, woolly and white beneath, and up to 10 cm (4 in.) long.

Origin Like wormwood, mugwort was first introduced to the east coast of North America from Eurasia in the early nineteenth century. European settlers brought it over to grow and use for its medicinal and culinary properties.

Etymology The genus name *Artemisia* commemorates the Greek goddess of chastity, Artemis (the Roman Diana).

Habitat Prefers full sun in moist or dry soils, roadsides, waste areas, forest edges—usually where there is (or was) human habitation.

Season Flowers from July to October.

Traditional/Medicinal Use Mugwort was highly valued in European apothecaries since medieval times. It was used for all manner of conditions, including digestive issues and menstrual complaints.

Edibility The raw leaves were used as one of the main ingredients for stuffing geese and as a flavouring for beer prior to hops. The dried leaves were crushed and used as a condiment or used as a potherb.

MULLEIN

GREAT MULLEIN / VELVET DOCK / TORCHES / CANDLESTICK / BEGGAR'S BLANKET

Verbascum thapsus

FIGWORT FAMILY Scrophulariaceae

Description Mullein is a taprooted biennial 1.2–2.4 m (4–8 ft.) in height. Its yellow flowers are up to 2.5 cm (1 in.) across, stalkless, and borne in elongated terminal spikes. The velvet-like basal leaves are 15–45 cm (6–18 in.) long, with woolly hairs. The second-year stem leaves are alternating and reduced in size upward. In the first year, mullein puts out a taproot and sits as a basal rosette of leaves. In the second year, it puts up a giant stem of leaves and flowers.

Origin Introduced to North America from Eurasia as a medicinal plant.

Etymology The genus name *Verbascum* is thought to be a corruption of the Latin word *barbascum*, meaning "with bread." The species name *thapsus* refers to an ancient town in what is now Tunisia; it may also have been named for the Greek island of Thapsos.

Habitat Usually accepted as a good-looking addition to the garden. Prefers full sunlight and is not fussy about soils.

Season Blooms mid-summer until autumn.

Traditional/Medicinal Use Mullein has long been used in medicine to treat respiratory diseases, diarrhea, gout, burns, ringworm, hemorrhoids, and warts. The leaves and stems were used to make candle wicks and shoe insoles, and were dipped in fat or wax for torches.

Edibility A strained tea can be made from the hairy leaves and flowers.

MUSK MALLOW

Malva moschata

MALLOW FAMILY Malvaceae

Description Musk mallow is a short-lived taprooted herbaceous perennial 60 cm (24 in.) in height. Its flowers are up to 5 cm (2 in.) across, have five petals, and range in colour from white to pink to purple. The leaves are up to 8 cm (3 in.) across and palmately compounded into several dissected linear segments.

Origin Introduced from Eurasia.

Etymology The well-known spongy candy marshmallow used to be made from the gummy roots of the closely allied marsh mallow (A. *officinalis*). The species name *moschata* means "musk-like," in reference to the musky scented leaves.

Habitat Roadsides, back alleys, abandoned lots, and fields.

Season Flowers from June to September.

Traditional/Medicinal Use Mallows have been used for centuries in poultices to treat bruises, ulcers, boils, sores, cuts, skin disorders, swelling, and inflammation.

Edibility The young leaves and flowers are best used raw in salads. The leaves become very gluey or slimy when cooked. The seeds, when eaten raw, have a nutty flavour.

NIPPLEWORT

Lapsana communis

ASTER FAMILY Asteraceae

Description Nipplewort is a taprooted annual 0.4–1.5 m (1.3–5 ft.) in height. Its yellow flowers are 2 cm (1 in.) across and composed of ray florets only, with no disc. The flowers last only a day or two, quickly followed by miniature dandelion-like seed heads. The lower leaves are egg shaped and coarsely toothed, with basal lobes. The upper leaves are reduced and narrow.

Origin Introduced from Eurasia.

Etymology The genus name *Lapsana* is from the Greek word *lapazo*, meaning "purge," indicating that the plant was used medicinally.

Habitat Open forests, forest edges, roadsides. Mainly an understory plant.

Season Flowers from June to September.

Traditional/Medicinal Use A decoction from nipplewort was also used as treatment for sore or cracked nipples, as indicated by its common name.

Edibility Throughout history, nipplewort has been cultivated in Europe as a vegetable. The young leaves can be used quite enjoyably as salad greens and potherbs.

NODDING ONION

Allium cernuum

LILY FAMILY Liliaceae

Description Nodding onion is an herbaceous perennial up to 45 cm (18 in.) in height. Over a dozen small pink flowers are held in the distinctive nodding umbels. The grassy leaves are basal, up to 30 cm (12 in.) long, and similar to those of a green onion. Both bulbs and leaves smell of onion. The district of Lillooet was once covered in nodding onions; in the Salish language, Lillooet means "place of many onions."

Etymology The species name *cernuum* means "nodding."

Habitat Dry grassy slopes, rocky outcrops, and forest edges at lower elevations.

Season Flowers from June to August.

Edibility The bulbs can be eaten raw or cooked. The greens, too, are edible.

ORACHE

Atriplex patula

GOOSEFOOT FAMILY Chenopodiaceae

Description Orache is a fleshy annual 0.3–1 m (1–3.3 ft.) long. Its leaves are variable; the upper leaves are lanceolate, toothed, and alternate, and the lower leaves are larger, more arrowhead shaped, opposite, and 5–8 cm (2–3 in.) long. The inconspicuous green flowers form in clusters, both terminal and in the axils of the leaves. The seeds and fruit are enclosed by two triangular bracts.

Habitat Saline soils along the coast.

Season In flower and fruit from July to October.

Edibility Orache is in the spinach family and is prepared in the same way as spinach, such as in sauces, stews, and as a potherb. As with other plants containing oxalate salts, moderation in consumption is recommended.

OREGON OXALIS

Oxalis oregana

WOODSORREL FAMILY Oxalidaceae

Description Oregon oxalis is a delicate-looking perennial up to 15 cm (6 in.) tall. The white to pinkish flowers are borne singly on slender stalks that are shorter than the leaves. The shamrock-like leaves have three heart-shaped leaflets up to 15 cm (6 in.) tall. Both the flower and leaf stalks arise from the base.

Etymology The genus name *Oxalis* means "acid," "sour," or "sharp," in reference to the taste of the leaves.

Habitat Moist forested areas at low to mid elevations.

Season Flowers from April to June.

Edibility The leaves can be eaten, though sparingly, for they contain oxalic acid. The stems, when collected in abundance, can be baked with sugar or honey, not unlike rhubarb.

PENNYCRESS

FIELD PENNYCRESS / MITHRIDATE MUSTARD / STINKWEED

Thlaspi arvense

MUSTARD FAMILY Brassicaceae

Description Pennycress is a taprooted annual 80 cm (32 in.) in height. Its tiny white flowers have four petals and are borne in compact terminal racemes that elongate as the seeds are produced. The seed pods (siliques) are 1.3–1.7 cm (0.5–0.7 in.) long, roundly heart shaped, and flattened, with a notch at the top. The basal leaves are formed in a rosette. They are larger and more oval than the alternate lanceolate upper leaves.

Origin Native to Asia; introduced from Europe in the early 1700s.

Etymology The common name pennycress is from the dried seed pods' resemblance to pennies. Crush the leaves of the plant and you will know why it's also called stinkweed.

Habitat Disturbed soils, abandoned fields, and gardens.

Season Flowers from April to August.

Traditional/Medicinal Use In the middle ages, pennycress was used in the complicated mixture mithridate, a mustard thought to be an antidote to poisons.

Edibility The young shoots and leaves can be eaten raw or cooked; they were typically eaten with other vegetables.

PERENNIAL SOW THISTLE

CORN SOW THISTLE / CREEPING SOW THISTLE

Sonchus arvensis

ASTER FAMILY Asteraceae

Description Three species and one variety of sow thistle are commonly seen in the Pacific Northwest: perennial sow thistle (*S. arvensis*) annual sow thistle (*S. asper*), common sow thistle (*S. oleraceus*), and hairless or smooth perennial sow thistle (*S. arvensis* var. *glabrescens*). They can be distinguished from true thistles by their milky juice (latex) and dandelion-like yellow flowers.

Perennial sow thistle is an herbaceous perennial up to 1.8 m (6 ft.) in height. Its flowers are 5 cm (2 in.) across, composed of ray florets only, and borne in round or flat-topped clusters. Floral bracts supporting the flower head that are not covered with woolly hairs are most likely of the smooth variety, *S. arvensis* var. *glabrescens*. The leaves are 5–40 cm (2–16 in.) long, clasping with prickly margins, and deeply lobed.

Origin Introduced from Europe to North America. The Romans may have brought sow thistle to England.

Habitat Construction sites more than a year old, vacant lots, roadsides, and boulevards that are not accustomed to lawnmowers.

Season Flowers from August to October.

Edibility Young leaves taste like dandelion and can be cooked like spinach.

PIGWEED RED ROOT

Amaranthus retroflexus

AMARANTH FAMILY Amaranthaceae

Description Pigweed is a an annual up to 1.8 m (6 ft.) in height, more commonly seen at 0.9–1.2 m (3–4 ft.). It has coarse foliage, and its small green flowers are borne in densely crowded spikes 5–15 cm (2–6 in.) long. The abundant seeds are small, oval, and glossy black. The mature leaves are long stemmed, wavy edged, and to 15 cm (6 in.) long. Other common species of *Amaranthus* are tumbling amaranth (*A. albus*) and prostrate pigweed (*A. graecizans*).

Origin Native to tropical America; introduced or spread to the West Coast by 1940.

Etymology The common name pigweed is from the pleasure hogs get from eating it. The genus name *Amaranthus* is from the Greek word *amarantos* "unfading," in reference to the flowers retaining their colour for a long time.

Habitat Enjoys deep moist fertile soils, like around farms and irrigated parks and golf courses.

Season Flowers from July to the first frost.

Traditional/Medicinal Use An extract from the plant was thought to be an antidote to snakebites.

Edibility The spring and early-summer leaves are excellent as potherbs or raw in salads.

PINEAPPLE WEED DISC MAYWEED

Matricaria discoidea

ASTER FAMILY Asteraceae

Description Pineapple weed is a pleasantly scented annual up to 30 cm (12 in.) in height. Its green-yellow flowers are rayless (all disc) and borne in dense terminal clusters. The alternating leaves are divided one to three times, giving the plant a fern-like appearance.

Origin Pineapple weed is native to North America. It was probably introduced to the West Coast.

Etymology The genus name *Matricaria* is medieval for "mother" or "womb." The species name *discoidea* means "disc only"; the flowers have no rays.

Habitat It has the ability to grow in compacted soils, usually on gravel roads and pathways, and in waste areas.

Season Flowers from June to August.

Traditional/Medicinal Use It was used in midwifery and to treat uterine infections. The pineapple scent from the crushed plants makes it useful for potpourri.

Edibility Young plants make a good substitute for chamomile in herbal tea.

PRICKLY LETTUCE WILD LETTUCE / OPIUM LETTUCE

Lactuca serriola

ASTER FAMILY Asteraceae

Syn. *Lactuca scariola*

Description Prickly lettuce is a taprooted annual or biennial 0.9–1.8 m (3–5 ft.) in height. Its yellow ray florets have no disc, are up to 2.5 cm (1 in.) across and are borne in terminal clusters. The alternating leaves are 5–30 cm (2–12 in.) long, clasping, bluish green, and usually deeply incised. Prickly lettuce is thought to be the ancestor to our modern-day garden varieties.

Origin Introduced from Europe.

Etymology The common name prickly lettuce is from the weak prickles on the under midrib and leaf margins.

Habitat Forest edges, waste sites, and cultivated fields.

Season Flowers from July to September.

Traditional/Medicinal Use The milky juice in the plant contains lactucarium, a drug used for its antispasmodic, digestive, narcotic, and sedative properties. When dried, it is like a weak opiate without being addictive or a problem to the digestive system.

Edibility The young leaves can be eaten raw or as a potherb, and the young shoots can be cooked like asparagus.

CAUTION Consume in moderation as large quantities can cause digestive problems.

PURSLANE

LITTLE HOGWEED / PIGWEED / PUSSELY / GREEN PURSLANE

Portulaca oleracea

PURSLANE FAMILY Portulacaceae

Syn. *P. neglecta*

Description Purslane is a prostrate succulent annual 50 cm (12 in.) long. Its showy yellow flowers are 1 cm (0.5 in.) across and are borne both in the axils and in terminal clusters. The succulent leaves are up to 2.5 cm (1 in.) long, alternating, wedge shaped, green above, and pale purple below. The flowers open on sunny days only, a common trait in the purslane family.

Origin Introduced from Eurasia; first seen on the east coast of North America in the 1670s.

Etymology The species name *oleracea* is given to plants that are vegetables and potherbs.

Habitat Prefers dry exposed areas, driveways, waste areas, and parks.

Season The seeds germinate when the summer heat is full. Flowers from August to September.

Traditional/Medicinal Use Medicinally, purslane has been known since the time of Hippocrates. It was used by both Theophrastus and Pedanius Dioscorides for its diuretic, anti-parasitic, and cathartic properties.

Edibility The leaves can be eaten raw or as a potherb. The seeds can be dried, ground, and mixed with flour.

QUICKWEED

SHAGGY GALINSOGA / FRINGED QUICKWEED / SHAGGY SOLDIER

Galinsoga quadriradiata

ASTER FAMILY Asteraceae

Description Quickweed is a fibrous-rooted herbaceous perennial 20–60 cm (8–24 in.) in height. Its flowers have yellow disc florets surrounded by white ray florets, which are borne in terminal leafy-bracted clusters. The leaves are 2.5–8 cm (1–3 in.) long, opposite, hairy, and coarsely toothed. Small-flowered quickweed (*G. parviflora*) is very similar, except it is not as hairy and its ray florets are slightly smaller.

Origin Introduced from Central and Southern America.

Etymology The genus name *Galinsoga* commemorates the eighteenth-century Spanish physician and botanist Don Mariano Martinez de Galinsoga (1756–97). The common name quickweed indicates how fast this plant can spread.

Habitat Likes cultivated fields, gardens, and fertile soils.

Season Flowers from May to December.

Edibility The fuzzy leaves can be eaten raw in salads or cooked as potherbs.

▶ RED CLOVER PURPLE CLOVER / MEADOW HONEYSUCKLE

Trifolium pratense

PEA FAMILY Fabaceae

Description Red clover is a well-recognized herbaceous perennial up to 60 cm (24 in.) in height. Its pink to red flower heads are 2.5 cm (1 in.) across, fragrant, and composed of fifty to two hundred pea-like florets. The florets open from the bottom of the head and progressively upward. The alternating leaves are pinnately divided into three leaflets (or, rarely, four), with some of the leaflets having inverted V-shaped stripes known as chevrons.

Similar Species White clover (*T. repens*) has smaller flowers and smaller, rounder leaflets.

Origin Introduced from Eurasia as a forage crop.

Etymology The common name clover is derived from the Latin word *cava*, "club" as in the "clover club" in playing cards and the three-pronged club used in combat by Hercules. The Latin name translates to "three-leaved plant from the meadows."

Habitat Cultivated fields, lawns, and boulevards not accustomed to the grooming of a lawnmower.

Season Flowers from May to November.

Traditional/Medicinal Use The dried flowers make a pleasant tea that may treat bronchial problems and whooping cough.

Edibility Clover leaves can be eaten raw in moderation. They are better suited as potherbs.

RIBWORT PLANTAIN NARROW-LEAVED PLANTAIN

Plantago lanceolata

PLANTAIN FAMILY Plantaginaceae

Description Ribwort plantain is an herbaceous perennial 15–50 cm (6–20 in.) in height. Its unusual flowers have four greenish petals and are borne in dense terminal clusters on leafless stalks. The leaves are 10–30 cm (4–12 in.) long, basal, and strongly ribbed.

Origin Introduced from Eurasia by the early settlers.

Etymology The species name *lanceolata* refers to the lance-shaped leaves.

Habitat Common on open ground to mid elevations wherever settlements are or have been.

Season Flowers from May to October.

Traditional/Medicinal Use A beautiful golden dye can be made by boiling the entire plant and letting the water sit for 24 hours. The leaves can be used externally to stop blood flow from cuts and wounds. Internally, it has been used to treat diarrhea, hemorrhoids, asthma, and gastritis. In the tropics, the crushed leaves are used to treat insect bites and stings. Capsules of plantain are available at most herbal stores.

Edibility Although bitter, the young leaves of both ribwort plantain and common plantain can be eaten raw or as potherbs. The seeds can be dried, ground, and added to flour when baking.

RUSSIAN ORACHE TWOSCALE SALTBUSH

Atriplex micrantha

GOOSEFOOT FAMILY Chenopodiaceae

Syn. *A. heterosperma*

Description Russian orache is a fleshy sprawling annual up to 1.5 m (5 ft.) in height. Its inconspicuous flowers are greenish and borne both in the leaf axils and in terminal spikes. The arrow-shaped leaves are 5 cm (2 in.) long, with the lower leaves opposite and the upper alternating. As with most oraches, the surfaces of the young leaves have a mealy white powder that disappears as they mature. The tiny seeds are borne in bumpy triangular bracts.

Origin Introduced from Eurasia.

Etymology The species name *micrantha* means "small flowered." Garden orache, or French spinach (A. *hortensis*), has the largest leaves of the oraches and has long been used as a spinach substitute in Europe.

Habitat Dry roadsides and waste areas.

Season Flowers from July to September.

Edibility Can be used the same way spinach is used. As with other plants containing oxalate salts, moderation in consumption is recommended.

SALSIFY PURPLE GOAT'S BEARD / OYSTER PLANT

Tragopogon porrifolius

ASTER FAMILY Asteraceae

Description Salsify is a taprooted biennial 0.6–1.2 m (2–4 ft.) in height. Its attractive purple flowers are 5–7 cm (2–3 in.) across with bright-green floral bracts extending beyond the ray florets. Each flower is borne singly at the end of a hollow stem. The long grass-like leaves are up to 30 cm (12 in.) long and clasp the stems at the base.

Origin Introduced from Europe.

Etymology The genus name *Tragopogon* is from the Greek, meaning "goat's beard." It was thought that the fluffy seed heads resembled a goat's beard as they fell apart. The species name *porrifolius* is Latin for "leaves resembling leeks."

Habitat I have found most of the two species of salsify on the San Juan and Gulf Islands and about 2,000 m (6,200 ft.) up our Coastal Mountains in grassy meadows.

Season Flowers from July to August.

Traditional/Medicinal Use Decoctions of the plant were used to treat liver and stomach disorders.

Edibility Salsify roots have been baked, sautéed, and used in soups and stews. The roots are said to taste like oysters (hence the common name oyster plant).

SEABEACH SANDWORT

Honckenya peploides

PINK FAMILY Caryophyllaceae

Description Seabeach sandwort is an herbaceous perennial up to 30 cm (12 in.) in height. Its odd-looking flowers are greenish white, up to 1.5 cm (0.75 in.) across, and held in terminal-leaf whorls. The fleshy leaves are elliptical, pale green, opposite, and up to 5 cm (2 in.) long. Sandwort can be seen growing as a single plant or as a mat up to 1 m (3.3 ft.) across.

Origin This is a circumboreal plant. It is found in coastal areas in Europe, Asia, and North America. I was exploring the shores of Iceland and came across vast stretches of it; the leaves and seeds were being collected for food by the locals.

Etymology The genus name refers to Gerhard August Honckeny (1724–1805), an eighteenth-century German botanist.

Habitat Upper sandy beaches, between rocks and logs.

Season Flowers from July to August.

Edibility The young shoots and leaves are high in vitamins A and C and can be eaten fresh or cooked. The seeds can be used as a garnish or ground up and added to flour.

SEA-WATCH ANGELICA

Angelica lucida

CARROT FAMILY Apiaceae

Description Sea-watch angelica is a taprooted, herbaceous perennial 1.4 m (4.6 ft.) tall. The small white flowers are held in small heads, which form a compound umbel up to 15 cm (6 in.) across. The hairless leaves are more rounded than kneeling angelica and lack the bend in the leaf stalk.

Etymology The common name of sea-watch angelica is a bit obscure. However, if you see the autumn stalks on the bluffs above the Pacific coastline, it looks as if they are keeping watch over the sea.

Habitat Mainly seen in moist seepage areas along the coast.

Season Flowers from July to August.

Edibility The stems can be peeled and eaten.

SHEEP SORREL

SOUR GRASS / WILD SORREL / RED SORREL

Rumex acetosella

BUCKWHEAT FAMILY Polygonaceae

Description Sheep sorrel is a rhizomatous, herbaceous perennial up to 40 cm (16 in.) in height. Its flowers are dioecious, with male and female flowers on separate plants. They are yellowish to red and borne in open terminal clusters. Being in the buckwheat family, the flowers have no petals. The mature leaves develop basal lobes, giving them an arrow-shaped appearance.

Origin Introduced from Eurasia.

Etymology Sheep sorrel is also known as sour grass because of its sharp taste, like oxalic acid. Both the genus name and species name refer to the sour-tasting leaves.

Habitat Loves acidic soils, which the coastal Pacific Northwest has in abundance. Because blueberries also thrive on acidic soil, blueberry farmers have a strong dislike of this plant. The heavy continual rainfall along our coast has made most of the soil slightly (to strongly) acidic (6.5–4.5 pH), which is the range sheep sorrel enjoys.

Season Flowers from April to October.

Traditional/Medicinal Use Sheep sorrel was used as blood cleanser and gargle and to treat fevers, sores, and ringworm.

Edibility The sharp-tasting leaves can be added to salads, soups, and stews.

SHEPHERD'S PURSE

SHEPHERD'S POUCH / PEPPER-AND-SALT / PICKPOCKET

Capsella bursa-pastoris

MUSTARD FAMILY Brassicaceae

Description Shepherd's purse is a taprooted annual 10–50 cm (4–20 in.) in height. Its tiny flowers are borne in tight clusters that thin out along the stem as the seeds are produced. The basal leaves are up to 12 cm (5 in.) long and pinnately lobed; they sit in a rosette. The stem leaves are smaller, sawtoothed, and clasping, with two round lobes at their base. The seed pods (siliques) are up to 0.7 cm (0.25 in.) long, heart shaped or triangular, and filled with tiny seeds.

Origin Native to Europe and Asia Minor.

Etymology The genus name *Capsella* means "small box." The species name is derived from the words *bursa* "purse" and *pastoris* "shepherd" in reference to the old leather purses shepherds wore on their belts.

Habitat Can be found almost anywhere along the coast from sea level to sub-alpine elevations. Anywhere humans have colonized, shepherd's purse will be close by.

Season Seeds and flowers can be seen at the same time throughout summer to autumn.

Traditional/Medicinal Use Medicinally, shepherd's purse was one of the most used plants in the mustard family. One of its main uses was to stop hemorrhaging of all kinds.

Edibility The plant has a bitter, peppery taste. The young leaves and seeds were used as potherbs.

SIBERIAN MINER'S LETTUCE

Claytonia sibirica

PURSLANE FAMILY Montiaceae

Description Siberian miner's lettuce is a small annual up to 30 cm (12 in.) tall. Its small white to pink flowers are five petalled and produced in abundance on long, thin, fleshy stems. The basal leaves are long stemmed, opposite, ovate, and, like the stems, succulent. Another species, miner's lettuce (*C. perfoliata*), differs in that its upper leaves are disc shaped and fused to other flower stems.

Etymology Siberian miner's lettuce was first discovered in Russia, where it was a staple food for miners and prospectors.

Habitat Moist forest areas at low to mid elevations. Siberian miner's lettuce prefers cool moist forest floors.

Season Flowers from mid-April to July, depending on elevation.

Traditional/Medicinal Use The leaves have been applied to cuts and sores and the squeezed juice used for eye drops. More recently it has been used as an antidandruff.

Edibility The leaves can be used both as a salad green and a potherb and are a good source of vitamin C.

SILVERWEED

Potentilla anserina ssp. *pacifica*

ROSE FAMILY Rosaceae

Description Silverweed grows to only 30 cm (12 in.) in height but can take over several hectares or acres in favourable conditions. The yellow flowers are produced singly on a leafless stalk. The compound leaves reach 25 cm (10 in.) in length and have nine to nineteen toothed leaflets. They are bi-coloured: grass green above and felty silver below (hence the common name). Silverweed spreads quickly thanks to its fast-growing stolons, which root at the nodes.

Etymology The genus name *Potentilla* means "powerful," a reference to its medicinal properties.

Habitat Saline marshes, meadows, and wet runoff areas near the ocean.

Season Flowering starts mid-May and continues through June and July.

Edibility The leaves can be cooked or used raw. Some thick concentrations of silverweed were so important to coastal peoples that chiefs demanded ownership of the patches. The longer roots were roasted and prized by coastal groups.

SKUNK CABBAGE

Lysichiton americanus

ARUM FAMILY Araceae

Description Skunk cabbage is an herbaceous perennial up to 1.5 m (5 ft.) in height and as much as 2 m (6.6 ft.) across. The small greenish flowers are densely packed on a fleshy spike and surrounded by a showy yellow spathe, the emergence of which is a sure sign that spring is near. The tropical-looking leaves can be over 1 m (3.3 ft.) long and 50 cm (20 in.) wide.

Habitat Common at lower elevations in wet areas such as springs, swamps, seepage areas, and floodplains.

Season Flowers from May to July.

Edibility The roots can be cooked and eaten. Skunk cabbage was an important traditional spring food in times of famine, and it is said that this poorly named plant has saved the lives of thousands.

CAUTION Do not eat any other parts of the plant raw or cooked, for they contain calcium oxalate, which will cause burning and swelling in the mouth.

▶ SPREADING STONECROP

Sedum divergens

STONECROP FAMILY Crassulaceae

Description Spreading stonecrop is a mat-forming perennial 10–15 cm (4–6 in.) in height. Its yellow flowers have five petals held in clusters by stems 10–15 cm (4–6 in.) tall. The green to red leaves are plump, round, succulent, and less than 1 cm (0.5 in.) across.

Etymology The genus name *Sedum* comes from the Latin, meaning "to sit," referring to the way the leaves are placed.

Habitat Rocky bluffs, exposed talus slopes at low to high elevations.

Season Flowers from June to August, depending on elevation.

Edibility The leaves can be eaten raw, in moderation, and were a traditional food of Haida people.

SPRINGBANK CLOVER

Trifolium wormskioldii

PEA FAMILY Fabaceae

Description Springbank clover is an herbaceous perennial that is usually prostrate up to 30 cm (12 in.) in height. The flowers, up to 3 cm (1 in.) across, are pink to reddish purple and often white-tipped. The finely toothed leaves are typically clover shaped, with three leaflets joined at one point.

Etymology The species name commemorates Danish botanist Morten Wormskjold (1783–1845).

Habitat Common along the coastline in moist areas from sea level to mid elevations.

Season Flowers from June to August.

Edibility The rhizomes can be dug up, cleaned, and steamed. Springbank clover was an important vegetable on the West Coast and often stored for winter use.

▶ STAR OF BETHLEHEM

STARFLOWER / SLEEPYDICK / NAP-AT-NOON / DOVE'S DUNG

Ornithogalum umbellatum

LILY FAMILY Liliaceae

Description Star of Bethlehem is a bulbous herbaceous perennial with flowering stems up to 30 cm (12 in.) in height. Its star-shaped flowers are bright white with outer green stripes and are borne in flat-topped clusters of five to twenty. The grass-like leaves are 15–30 cm (6–12 in.) long and basal, with a thin white stripe down the keel.

Origin Introduced from Europe as a garden ornamental.

Etymology The genus name is derived from the Greek words *orni*, "bird," and *gala*, "milk." The bulbs are thought to be the "dove's dung" mentioned in the Bible, which were sold during the Babylonian siege of Jerusalem. The flowers of star of Bethlehem close at night and reopen when there is warmth and sun—they will not open on cold cloudy days.

Habitat Has jumped the garden fence and can be seen at lower elevations where humans have settled.

Season Flowers from April to June; then the leaves wither and disappear.

Traditional/Medicinal Use Modern herbal stores sell tinctures of the plant with a twenty-seven percent alcohol content for relief of naturally occurring nervous tension. I wonder if the alcohol has any effect.

Edibility In ancient Europe, the bulbs were eaten raw, boiled, or roasted like chestnuts; however, today, some authorities say the bulbs are poisonous to humans and livestock.

▶ STAR-FLOWERED SOLOMON'S SEAL

Maianthemum stellatum

LILY FAMILY *Liliaceae*

Description Star-flowered Solomon's seal is a smaller, more refined plant than false Solomon's seal (*S. racemosa*). It grows to a height of 60–70 cm (24–28 in.) and has attractive white star-shaped flowers that grow in open terminal clusters. The broad lance-shaped leaves grow on short stalks and are alternate and 15 cm (6 in.) long; they are usually folded down the midrib and have somewhat clasping bases. The round immature fruit is green with dark stripes; it ripens slowly, turning dark bluish black. The star-shaped flowers and striped fruit distinguish this species from other *Maianthemum* spp.

Etymology The species name *stellatum* references the star-shaped flowers.

Habitat Moist shaded forests, often in association with devil's club (*O. horridus*) at low to mid elevations.

Season Flowers from May to June. Berries are seen by mid-August.

Edibility While the berries do not have much flavour, they are edible.

STINGING NETTLE

Urtica dioica

NETTLE FAMILY Urticaceae

Description Stinging nettle is an herbaceous perennial growing to over 2 m (6.6 ft.) high. Its tiny flowers are greenish and produced in hanging clusters up to 5 cm (2 in.) long. The leaves are heart shaped at the base, tapered to the top, coarsely toothed, and up to 10 cm (4 in.) long.

Etymology The genus name *Urtica* is from the Latin word *uro* "to burn."

Habitat Thrives in moist, nutrient-rich, somewhat shady disturbed sites, where it can form great masses. Stinging nettles are usually an indicator of nitrogen-rich soil.

Season Flowering starts at the beginning of May.

Traditional/Medicinal Use The root was often used a diuretic.

Edibility The young leaves and new shoots up to 20 cm (8 in.) long can be boiled as spinach or used as potherbs.

CAUTION The stalks, stems, and leaves all have stinging hairs that contain formic acid; many people have the misfortune of encountering this plant the hard way.

STREAM VIOLET

Viola glabella

VIOLET FAMILY Violaceae

Description Stream violet is an herbaceous woodland perennial up to 25 cm (10 in.) in height. Its showy yellow flowers, up to 2 cm (1 in.) across, each have five petals; the top two petals are often pure yellow, while the bottom three have purple lines. The heart-shaped leaves are toothed and grow up to 5 cm (2 in.) across.

Habitat Needs a moist environment, such as open forests and meadows, at low to high elevations. Common at all elevations.

Season Flowers mid-May at lower elevations and mid-July at higher elevations.

Edibility The flower buds and young leaves can be used in salads or steeped for tea.

▶ TIGER LILY

Lilium columbianum

LILY FAMILY Liliaceae

Description Tiger lily is an elegant herbaceous perennial up to 1.5 m (5 ft.) tall. Its drooping flowers go from deep yellow to bright orange. A vigorous plant can have twenty or more flowers. Shortly after the flower buds have opened, the petals curve backward to reveal maroon spots and anthers. The leaves are lance shaped, usually in a whorl, and 5–10 cm (2–4 in.) long. It is said that anyone who smells a tiger lily will develop freckles.

Habitat A diverse range, including open forests, meadows, rock outcrops, and the sides of logging roads, at low to subalpine elevations.

Season Flowering starts mid-May.

Edibility The bulbs can be boiled or steamed.

WALL LETTUCE

Lactuca muralis

ASTER FAMILY Asteraceae

Description Wall lettuce is a taprooted annual or biennial 0.3–1.2 m (1–4 ft.) in height. Its yellow flowers are 2 cm (1 in.) across, have five petals, and are borne in terminal clusters. The basal leaves are 5–20 cm (2–8 in.) long, are deeply cut and lobed, clasp the stem, and are glaucous underneath. The few upper leaves are reduced in size.

Origin Wall lettuce is native to Europe.

Etymology The genus name *Lactuca* is descriptive of the milky juice "lac" that comes from the broken stems and leaves. The species name *muralis* refers to where the plant commonly grows on or by the garden wall.

Habitat As suggested in the etymology, wall lettuce commonly grows on stone garden walls.

Season Flowers from July to September.

Edibility The young leaves have long been used fresh in salads and as potherbs. Boiling the leaves removes some of the bitterness.

▶ WAPATO ARROWHEAD / DUCK POTATO

Sagittaria latifolia

WATER PLANTAIN FAMILY Alismataceae

Description Wapato is an herbaceous freshwater perennial that can grow up to 90 cm (36 in.) tall. The arrow-shaped leaves, up to 25 cm (10 in.) long, grow on long, slightly arching stalks. The waxy white flowers are produced in whorls of three on long leafless stems.

Etymology The common name wapato is from Chinook, meaning "tuberous plant." When wapato leaves and small tubers can be seen floating on the water, it usually is an indication of ducks or muskrats that have dislodged the plants for their starchy tubers (hence the name "duck potato").

Habitat Low elevations, shallow ponds, sloughs, lake edges, and slow-moving streams.

Season Depending on location, it blooms from June to early July.

Edibility The starchy tubers can be eaten raw or cooked and treated very much like potatoes; they can also be made into a flour. Wapato tubers were an important traditional food source and trading item.

WATER PARSNIP

Sium suave

CARROT FAMILY Apiaceae

Description Water parsnip is a lanky perennial that grows to 0.5–1.5 m (1.7–5 ft.) in height. The tiny white flowers, 0.3 cm (0.1 in.) across, are borne in masses resembling the spokes of an umbrella. The seven to fifteen finely toothed leaflets are supported by sheathed stalks.

Habitat Sloughs, lake, and pond edges at low to mid elevations.

Season Flowers from July to August.

Edibility The roots can be eaten raw or cooked and were considered an important traditional food source.

CAUTION Be careful when identifying this plant: many look-alikes are extremely poisonous. If you think you have found water parsnip, please consult a more detailed book to clearly identify the leaves and basal stems, e.g. *Vascular Plants of the Pacific Northwest* by Hitchcock, Cronquist, Ownbey, and Thompson.

WESTERN DOCK

Rumex occidentalis

BUCKWHEAT FAMILY Polygonaceae

Description Western dock is a strongly taprooted perennial 8 cm (3 in.) in height. The greenish flowers are held in dense clusters on the upper branched stems. The numerous basal leaves are heart to lance shaped with a heart-shaped to square base.

Etymology The genus name *Rumex* means "sour."

Habitat Damp or wet conditions, in fields and along coastlines, at low to mid elevations.

Season Flowers from June to July, with seeds persisting until September.

Traditional/Medicinal Use The leaves when crushed help combat the sting of stinging nettles.

Edibility Our native dock has edible young leaves as do any of the 200-plus species of docks around the world. This genus should be researched more as a potherb and raw vegetable.

▶ WESTERN SPRING BEAUTY

Claytonia lanceolata

PURSLANE FAMILY Montiaceae

Description Western spring beauty is an herbaceous perennial from a marble-sized corm up to 10–20 cm (4–8 in.) tall. The beautiful white to pink flowers have five petals with darker-pink veins.

Etymology The genus name *Claytonia* commemorates John Clayton (1686–1773). Clayton has been described as one of the greatest botanists in America. He corresponded with some of the greats of the day: George Washington, Thomas Jefferson, Carl Linnaeus, and John Bartram.

Habitat Mid to high elevations; usually seen chasing the snowpack as it melts.

Season As its common name suggests, flowering is usually from May to June.

Edibility The corms can be eaten raw or cooked and, traditionally, were often stored for winter food.

WHITE SWEET CLOVER

WHITE MELILOT / HONEY LOTUS / KING'S CLOVER

Melilotus alba

PEA FAMILY Fabaceae

Description White sweet clover is a taprooted annual or biennial up to 1.8 m (6 ft.) in height. Its fragrant white flowers are borne in elongated racemes 5–15 cm (2–6 in.) long that arise from the leaf axils. The leaves are alternating, veined, and typically clover-like, with three leaflets.

Origin Introduced from Eurasia in the 1600s as forage crops and bee plants.

Etymology The genus name *Melilotus* means "honey lotus," in reference to the amount of honey produced by the flowers.

Habitat Wastelands, dry roadsides, abandoned fields.

Season Flowers from May to August.

Traditional/Medicinal Use Plaster made from the plants was used to treat tumours and swelled joints. It was also used as an emollient and digestive. The flowers were used in a salve for skin sores.

Edibility The flowers and leaves can be used to make a sweet tea.

WILD BUCKWHEAT

BLACK BINDWEED / CLIMBING BUCKWHEAT / CORNBIND

Polygonum convolvulus

BUCKWHEAT FAMILY Polygonaceae

Syn. *Bilderdykia convolvulus, Fallopia convolvulus, Tinaria convolvulus*

Description Wild buckwheat is a sprawling and twining annual up to 1.2 m (4 ft.) in length. Its tiny flowers are greenish and borne both in the leaf axils and in terminal racemes. The black seeds are borne singly in three-sided capsules. The leaves are up to 5 cm (2 in.) long, alternating, and shaped like large arrowheads.

Similar Species When not in flower, wild buckwheat can be confused with field bindweed (*C. arvensis*).

Origin Introduced from Eurasia, it is thought to have come over to the West Coast in the 1860s.

Etymology The species name *convolvulus* means "leaves like bindweed."

Habitat Cultivated fields, landfills, abandoned property.

Season Flowers from May to October or until the first frost.

Edibility The seeds can be dried, ground, and mixed with flour.

WILD CARROT QUEEN ANNE'S LACE

Daucus carota

CARROT FAMILY Apiaceae

Description Wild carrot is a biennial up to 1 m (3.3 ft.) in height. Its small white flowers are grouped to form showy terminal clusters up to 10 cm (4 in.) across. The leaves are up to 15 cm (6 in.) long and dissected to the point that they resemble delicate ferns. If the stems are scratched, a carrot scent is released.

Origin Native to Europe; first recorded in the United States in 1739.

Etymology Both the genus and species name are Latin for "carrot." Although wild carrot was first introduced to England when Queen Elizabeth I was in reign (1533–1603), it is said that Queen Anne, who reigned from 1702 to 1714, was very fond of wearing lace, and it was named in her honour.

Habitat Ditches, boulevards, anywhere there is vacant land at lower elevations. I have photographed millions of these plants growing on large vacant properties on the Greek island of Mykonos.

Season Flowers from July to September.

Traditional/Medicinal Use Medicinally, a poultice from the roots was used to ease the pain of cancerous ulcers.

Edibility Wild carrot is thought to be the parent of our present-day cultivated carrot. The seeds of wild carrot can be used to spice up soups and stews.

WILD CHERVIL

Anthriscus sylvestris

CARROT FAMILY Apiaceae

Description Wild chervil is a taprooted perennial up to 1.2 m (4 ft.) in height. Its small white flowers are borne in loose umbrella-like clusters in the top portion of the plant. The leaves are triangular and finely dissected within. Care must be taken not to confuse wild chervil with the similar looking but extremely poisonous poison hemlock (*C. maculatum*).

Origin Wild chervil was first introduced to the east coast of North America. It is a fairly recent alien to the West Coast and is spreading rapidly.

Etymology The species name *sylvestris* means "growing in woods or forests," which it does very well.

Habitat Moist rich soils with partial shade on the edge of forested areas.

Season The plant is in flower from April to June, with seeds maturing from July onward. The stems die back in late summer, but a second growth of non-flowering stems and leaves appears in autumn and remains green throughout the winter.

Edibility The taproots are considered edible when boiled like parsnips.

CAUTION Do not confuse wild chervil with poison hemlock (*C. maculatum*), which looks similar but is extremely poisonous.

WILD GINGER

Asarum caudatum

BIRTHWORT FAMILY Aristolochiaceae

Description Wild ginger is a trailing evergreen perennial that forms patches several metres or feet wide. Its bell-shaped solitary flower is purplish brown, up to 5 cm (2 in.) across, with three pointed lobes. The heart-shaped leaves, up to 10 cm (4 in.) across, are formed in opposite pairs in the nodes. The whole plant has a mild ginger fragrance when crushed.

Habitat Moist shaded forests with rich humus, at low to mid elevations.

Season Flowers from mid-May to August.

Traditional/Medicinal Use The scented plants were put in bathwater, and the roots were boiled and drunk as a tea to ease stomach problems.

Edibility Roots can be cleaned and eaten raw or used as a ginger substitute.

WILD PROSO MILLET

MILLET / BROOM CORN MILLET / PANIC MILLET / PROSO MILLET

Panicum miliaceum

GRASS FAMILY Poaceae

Description Wild proso millet is a fibrous-rooted annual grass that grows up to 30 cm (12 in.) in soils and up to 1.2–1.8 m (4–6 ft.) in irrigated sites. Its flowers are in branched panicles 10–30 cm (4–12 in.) long, with the lower florets being sterile and the upper fertile. The leaves are up to 30 cm (12 in.) long, 2.5 cm (1 in.) wide at the base, and smooth to sparsely hairy. The seeds are smooth and shiny dark brown to black.

Origin Introduced from Eurasia as a grain and forage crop.

Habitat Often found around duck ponds and growing beneath bird feeders.

Season Flowers from July to September, with the seeds maturing from August to October.

Edibility It is extensively grown in Europe for cereals, flours, and poultry feed. In North America, it is mainly grown for birdseed, which explains where it is typically found. Millet is thought to be the very first of the cultivated grains.

WILD RADISH JOINTED CHARLOCK / WILD RAPE / WILD KALE

Raphanus raphanistrum / Raphanus sativus

MUSTARD FAMILY Brassicaceae

Description Wild radish is a thick taprooted annual or biennial up to 90 cm (3 ft.) in height. Its showy flowers are usually pale yellow, fading to white, with purple venation. The seed pods (siliques) are up to 8 cm (3 in.) long, with four to twelve seeds. The basal leaves are 10–20 cm (4–8 in.) long, pinnately divided, and hairy, with the stem leaves being reduced in size upward.

Origin Introduced from Europe. Cultivated radishes were grown by the Ancient Egyptians as early as 2700 BC.

Etymology The common name jointed charlock refers to the jointed compartments on the seed pods.

Habitat Moist soils in open areas, vacant lots, abandoned fields, and natural parks that don't employ lawnmowers.

Season Flowers from June to October.

Edibility The roots, cleaned but not peeled, can be eaten raw when small or steamed when larger.

CAUTION Seeds contain glucosinolates; avoid eating in large quantities.

WILD SARSAPARILLA

Aralia nudicaulis

GINSENG FAMILY Araliaceae

Description Wild sarsaparilla is an herbaceous perennial up to 40 cm (16 in.) in height. The easily overlooked flowers are greenish white, five petalled, and held in small rounded clusters; they are replaced by small clusters of dark-purple berries in August. Sarsaparilla produces from its rhizomes a central stem that has three compound leaves divided again into three to five leaflets.

Habitat Moist forests at low to mid elevations.

Season Flowers in June; ripe seed by August.

Edibility Young shoots can be used as a potherb, and the roots have been used as substitutes for true sarsaparilla (*Smilax* sp.) both in herbal medicine and as an uplifting drink.

▶ WILD TURNIP WINTER RAPE

Brassica napus

MUSTARD FAMILY Brassicaceae

Description Wild turnip is a taprooted annual up to 1.8 m (6 ft.) in height. The flowers are typical of the mustard family: dull yellow to golden, with four petals and four sepals. The seed pods (siliques) are 5–10 cm (2–4 in.) long. The leaves are alternating, heavily divided, and mostly clasping.

Origin All six species were introduced from Eurasia.

Etymology The genus name *Brassica* was the name given to cabbage-like plants by Pliny the Elder (23–79 AD), and the species name *napus* was Pliny's name for a turnip.

Habitat Cultivated gardens with moist soils, open areas, vacant lots, abandoned fields, and natural parks.

Season Flowers from April to July.

Edibility The young leaves, like those of its related species, can be eaten raw in salads or used as potherbs in soups and stews. Dried seeds ground and mixed with vinegar make a mustard paste.

WORMWOOD ABSINTHIUM / VERMOUTH

Artemisia absinthium

ASTER FAMILY Asteraceae

Description Wormwood is an aromatic herbaceous perennial up to 1.5 m (5 ft.) in height. Its small nodding flowers are roundish, green yellow, and borne in panicles in the upper leaf axils. The lower leaves are long stacked, dissected two to three times, and up to 10 cm (4 in.) long. As the leaves progress up the stems, they become shorter and less divided, until they are reduced to approximately 2.5 cm (1 in.) long and stalkless.

Origin Introduced to the east coast of North America from Eurasia in the early nineteenth century as a garden plant.

Etymology The oil extracted from wormwood is absinthol, which is the main ingredient in absinthe. Wormwood is reputed to have grown along the path of the serpent in the Garden of Eden.

Season Flowers from July to September.

Habitat Dry to moist soils in abandoned fields, roadsides, and upper beach areas.

Traditional/Medicinal Use Wormwood has been used medicinally for hundreds of years, including to relieve depression, jaundice, gout, and digestive problems. It has also been used as a worm expeller and moth repellant.

Edibility Wormwood is one of the main ingredients in the Swiss alcohol drink absinthe (see etymology). The French made this drink popular in the eighteenth century.

YARROW

Achillea millefolium

ASTER FAMILY Asteraceae

Description Yarrow is an herbaceous perennial up to 1 m (3.3 ft.) in height. Its many small white flowers form flat-topped clusters 5–10 cm (2–4 in.) across. The aromatic leaves are so finely dissected that they appear fern-like.

Origin There is still discussion as to whether early European settlers brought yarrow over to the new world or it was already here.

Etymology Its species name *millefolium* means "a thousand leaves." The genus name *Achillea* is from the Greek general Achilles, who is said to have used yarrow to stop the bleeding of his soldiers' wounds.

Habitat A common plant that can be found along untended boulevards, roadsides, and in vacant lots.

Season Flowers from July to August at higher elevations.

Traditional/Medicinal Use Herbal shops sell tablets, capsules, liquid formulas, and tinctures claiming relief from nose bleeds, fevers, colds, sore throats, high blood pressure, premenstrual syndrome, rheumatism, and a lot more, as well as aromatherapy. Yarrow's primary use since ancient times is for blood clotting.

Edibility Though bitter, the leaves can be used in salads. Also, a tea can be made from both the flowers and leaves.

YELLOW ARCHANGEL GOLDEN DEAD NETTLE

Lamiastrum galeobdolon

MINT FAMILY Lamiaceae

Description Yellow archangel is an evergreen perennial groundcover capable of covering hundreds of square metres or feet. Its helmet-shaped flowers are 1 cm (0.5 in.) long, yellow, and borne in whorls. The flowers have a hooded upper petal and a lipped lower petal. The lower lip serves as a landing pad for insects, and the reddish-brown markings act as nectar guides. The silver-green leaves are 5 cm (2 in.) long and ovate, with serrate edges.

Origin Introduced from Eurasia as a garden ornamental.

Etymology The common name yellow archangel is from the colour of the flowers and because it is in flower on May 8, the day commemorating the archangel Michael. The genus name *Lamiastrum* refers to its resemblance to plants in the genus *Lamium*.

Habitat Quite a few plants seen growing at the forest edge are from city property owners dumping their lawn and plant cuttings there the cuttings root and take off.

Season Flowers from April to June.

Traditional/Medicinal Use Medicinally, it was used to treat gout and joint pain, to stop bleeding, and to heal sores and ulcers.

Edibility The young shoots and leaves can be used in salads or as potherbs.

YELLOW GLACIER LILY

Erythronium grandiflorum

LILY FAMILY Liliaceae

Description Yellow glacier lily is an herbaceous perennial up to 30 cm (12 in.) in height. You will have to go high into the mountains to see the lily's bright-yellow flowers. The 20 cm (8 in.) long leaves are not mottled and are in pairs clasping the back stem. Yellow glacier lily can be seen by retreating or melting snowpacks, sometimes in the thousands.

Habitat Open moist slopes at mid to subalpine elevations.

Season Flowers from June to July at high elevations.

Edibility The bulbs, or corms, can eaten raw or steamed.

YELLOW POND LILY

Nuphar polysepalum

WATER LILY FAMILY Nymphaeaceae

Description Yellow pond lily is a long-stemmed aquatic perennial. Its striking yellow flowers, up to 10 cm (4 in.) across, are a familiar summer sight in lakes and ponds. A large round stigma dominates the centre of these large, waxy flowers. The heart-shaped floating leaves, or pads, grow to 40 cm (16 in.) long. The huge rhizomes when exposed at low water levels are sought after by bears.

Etymology The genus name *Nuphar* means "water lily."

Habitat Ponds, lakes, and marshes at low to mid elevations.

Season Flowering begins in May and continues through the summer.

Traditional/Medicinal Use The rhizomes/roots were roasted and used to treat tuberculosis.

Edibility The seeds are edible and can be ground into a flour or popped like popcorn; called *wokas* by Indigenous groups in California and Oregon, they were an important traditional food.

YELLOW SALSIFY GOAT'S BEARD / JACK-GO-TO-BED-AT-NOON

Tragopogon dubius

ASTER FAMILY Asteraceae

Description Yellow salsify is a taprooted biennial or, occasionally, short-lived perennial 30–90 cm (1–3 ft.) in height. Its yellow flowers are 7 cm (3 in.) across, with pointed floral bracts extending past the ray florets. The flower heads are borne singly at the end of hollow stems. The grass-like leaves are 10–30 cm (4–12 in.) long, bluish green, and clasping. The seed heads are much like a dandelion's except larger, up to 10 cm (4 in.) across.

Origin Introduced from Europe.

Etymology The flowers open at dawn and close by mid-afternoon; hence their common name Jack-go-to-bed-at-noon.

Habitat I have found most of the two species of salsify on the San Juan and Gulf Islands and about 2,000 m (6,200 ft.) up our Coastal Mountains in grassy meadows.

Season Flowers from July to September.

Traditional/Medicinal Use The milky juice from the plant was taken internally for stomach aches and heartburn.

Edibility The roots can be roasted or boiled in the same fashion as parsnips. The young stems collected before the flowers emerge can be cut into pieces and boiled like asparagus.

YELLOW SORREL

CREEPING WOODSORREL / CREEPING OXALIS / SOUR CLOVER / YELLOW WOODSORREL

Oxalis corniculata

WOODSORREL FAMILY Oxalidaceae

Syn. *Oxalis repens, Xanthoxalis corniculata*

Description Yellow sorrel is a taprooted herbaceous perennial up to 10 cm (4 in.) in height. Tiny yellow flowers are borne in terminal clusters of two to five on long slender stalks. Sticky brown seeds are borne in greyish capsules up to 2.5 cm (1 in.) long, which explode at maturity. The clover-like leaves are often reddish purple and have three heart-shaped leaflets that fold toward each other. In the evening or in cold weather, the leaf stalks droop and the leaves close.

Origin Introduced from Eurasia in the 1700s.

Etymology The genus name *Oxalis* is from the Greek word *oxys*, meaning "acid," "sour," or "sharp," in reference to the taste of the leaves.

Habitat Likes to have its roots under rocks, bricks, patio pavers at low elevations. Can be a pest in the greenhouse, on patios. Will take root in sidewalk cracks and low stone walls.

Season Flowers from June to September.

Traditional/Medicinal Use The whole plant was used to treat scurvy, urinary tract infections, and diarrhea.

Edibility The leaves can be eaten raw or as a potherb but should not be eaten in large quantities as they contain oxalic acid.

YERBA BUENA

Satureja douglasii

MINT FAMILY Lamiaceae

Description Yerba buena is a fragrant trailing herbaceous perennial up to 1 m (3.3 ft.) long. Its inconspicuous flowers are white or slightly purple and borne in the leaf axils. The egg-shaped leaves grow opposite each other up to 3 cm (1 in.) long and are bluntly toothed and scented when crushed.

Etymology The common name yerba buena, meaning "good herb," was given to this plant by missionary Spanish priests in California.

Habitat Dry open forests.

Season Flowers from June to July.

Edibility The leaves can be steeped for a refreshing tea.

BERRIES

▶ BLACK GOOSEBERRY

Ribes lacustre

CURRANT AND GOOSEBERRY FAMILY Grossulariaceae

Description Black gooseberry is an armed shrub 2 m (6.6 ft.) in height. Its delicate reddish flowers are disc shaped and 0.7 cm (0.25 in.) across and hang in drooping clusters of seven to fifteen. The small, dark-purple berries are bristly and hang in clusters of three to four. The leaves are maple shaped, with five lobes 5 cm (2 in.) across. The branches are covered with small golden spines, with larger spines at the nodes.

Habitat Moist open forests and lake edges at low to high elevations.

Season The berries ripen from July to August.

Edibility The berries are edible. Traditionally the berries were eaten fresh or dried for winter use.

CAUTION Use caution when picking the berries: the spines can cause an allergic reaction in some people.

BLACK RASPBERRY BLACKCAP

Rubus leucodermis

ROSE FAMILY Rosaceae

Description Black raspberry is an armed deciduous shrub 2 m (6.6 ft.) in height. Its white flowers are small, up to 3 cm (1 in.) across and borne in terminal clusters of three to seven. The resulting fruit, 1 cm (0.5 in.) across, starts off red but turns dark purple to black by July or August. The leaves have three to five leaflets supported on long, arching, well-armed stems. Black raspberries can be distinguished from other raspberries by the bloom, a whitish waxy coating, on the stems.

Habitat Open forests and forest edges at low to mid elevations.

Season Flowers in June. The berries ripen from July to August, depending on elevation.

Edibility The berries are edible; traditionally they were eaten raw or dried into cakes for winter consumption.

▶ BOG BLUEBERRY

Vaccinium uliginosum

HEATHER FAMILY Ericaceae

Description Bog blueberry is a small deciduous bush 60 cm (24 in.) in height. At high elevations, it may reach only 10 cm (4 in.) in height. The tiny pink flowers give way to dark-blue berries that have a waxy coating. The berries are delicious. The leaves are green above and pale on the underside, up to 3 cm (1 in.) long, with no teeth.

Habitat Low elevated bogs along the coast to subalpine scrub.

Season The berries are ripe by July in lower areas.

Edibility The berries are edible and were traditionally eaten fresh or dried into cakes for winter use.

COASTAL STRAWBERRY

Fragaria chiloensis

ROSE FAMILY Rosaceae

Description Coastal strawberry is a deciduous maritime perennial that grows to 30 cm (12 in.) in height. Its white flowers are five petalled and up to 3 cm (1 in.) across, with approximately twenty orange stamens. The red fruit is small, 1.5 cm (0.75 in.) across, and tasty. The leathery leaves are basal and divided into three leaflets that are obovate, coarsely toothed, and woolly beneath.

Etymology The species name *chiloensis* refers to this plant's extensive range, from Alaska to Chile.

Habitat Usually seen on exposed rocky outcrops or in sand near the ocean.

Season Flowers from June to July.

Edibility The berries are edible, and the leaves can be steeped to make a refreshing tea.

CREEPING RASPBERRY FIVE-LEAVED BRAMBLE

Rubus pedatus

ROSE FAMILY Rosaceae

Description Creeping raspberry is an unarmed trailing perennial 0.2–1 m (0.7–3.3 ft.) long. Its white flowers, up to 2 cm (1 in.) across, are produced singly. The small edible berries are glossy red and form in clusters of one to five. The compound leaves have five coarsely toothed leaflets up to 2.5 cm (1 in.) across.

Habitat Found mostly at mid elevations in moist coniferous forests but can also be seen at low to subalpine elevations.

Season Flowers in June. The berries ripen from August to September.

Edibility The berries are edible and were traditionally eaten as they were picked but were not widely used because they are small and difficult to gather.

▶ CROWBERRY

Empetrum nigrum

CROWBERRY FAMILY Empetraceae

Description Crowberry is a low heather-like shrub 30 cm (12 in.) in height. The small purplish flowers are borne two to three in the leaf axils. The plants are mostly dioecious, with male and female flowers on separate plants. The crow-black berries are up to 1 cm (0.5 in.) across.

Habitat Bogs and bluffs along the coast.

Season Flowers from April to June, with the berries ripening from August to October, depending on elevation. I can remember going out in the beginning of October to pick them for Thanksgiving stuffing.

Edibility The berries are edible but, some would say, not particularly palatable. Crowberry was an important food source for the Inuit but not so highly regarded along the coast.

CUTLEAF BLACKBERRY

Rubus laciniatus

ROSE FAMILY Rosaceae

Description Cutleaf blackberry is a sprawling vine up to 6 m (20 ft.) in length. As its species name suggests the leaves are dissected, more so than the Himalayan species, and are palmately compounded with five leaflets. The fruit is up to 2.5 cm (1 in.) long and cherished by birds and smaller animals.

Origin Native to Eurasia.

Habitat Found at forest edges, in ditches, and on abandoned farmland.

Season Blooming starts mid-June, and the fruit sets by mid-August.

Edibility The unique tasting berries have been collected/harvested by humans for hundreds of years.

EVERGREEN HUCKLEBERRY

Vaccinium ovatum

HEATHER FAMILY Ericaceae

Description Evergreen huckleberry is an attractive mid-sized shrub up to 3 m (10 ft.) in height. In spring, it is covered in clusters of bell-shaped pinkish flowers. By late summer, the branches are weighed down by the many small blue-black berries, 0.7 cm (0.25 in.) across. This is a favourite late-season bush among avid berry pickers.

Habitat Coniferous forests at low elevations along the coast.

Season Flowering starts early May, and the berries ripen from early September to December.

Edibility The berries are edible. Traditionally the late-producing berries were in high demand for their flavour.

▶ GUMMY GOOSEBERRY STICKY GOOSEBERRY

Ribes lobbii

CURRANT AND GOOSEBERRY FAMILY Grossulariaceae

Description Gummy gooseberry is a spiny-stemmed shrub 1.2 m (4 ft.) in height. Its attractive flowers make it stand out in spring. The flowers have showy large recurved red sepals with white petals and stamens hanging down. The large berries are hairy; they start off green and ripen to purple. The leaves are small, with three to five lobes; the leaves and stems are sticky.

Etymology The species name *lobbii* refers to William Lobb (1809–64), who collected plants in North and South America for Veitch Nurseries.

Habitat Open forests along the coast at low to mid elevations.

Season The berries ripen by July.

Edibility The berries are edible (if not particularly palatable) and were traditionally eaten only in small quantities.

▸ HIGHBUSH CRANBERRY

Viburnum edule

HONEYSUCKLE FAMILY Caprifoliaceae

Description Highbush cranberry is usually seen as a straggling bush 3 m (10 ft.) in height. Its small white flowers, up to 1 cm (0.5 in.) across, are borne in rounded clusters nestled between the paired leaves. The resulting red berries, up to 1.5 cm (0.75 in.) across, grow in clusters of two to five. The leaves are opposite and mostly three-lobed.

Habitat Open forests and forest edges at low to mid elevations.

Season Flowers from the end of May to June. The berries ripen in September and last through October.

Edibility The tart berries are edible and were traditionally preserved for several months before eating.

HIMALAYAN BLACKBERRY ARMENIAN BLACKBERRY

Rubus discolor

ROSE FAMILY Rosaceae

Description Himalayan blackberry, now called Armenian Blackberry, is a well-armed vine up to 9m (30 ft.) across. Its berries are well sought after by novice and serious berry pickers.

Origin This blackberry is native to Armenia and Iran and was introduced to North America in 1885.

Habitat Far more common than the cutleaf blackberry, can be seen in both damp and dry soils in fields, parks, meadows, stream and lake sides at lower elevations.

Season Blooming starts mid-June, and the fruit sets from end of July to September.

Edibility When fully ripe the berries are delicious and have been the summer diet of many people and animals.

▸ OREGON GRAPE

Mahonia nervosa

BARBERRY FAMILY Berberidaceae

Description Oregon grape is a small spreading understory shrub that is very noticeable when the upright bright-yellow flowers are in bloom. By mid-summer, the clusters of small green fruit, 1 cm (0.5 in.) across, turn an attractive grape blue. The leaves are evergreen, holly-like, waxy, and compound, usually with nine to seventeen leaflets. The bark is rough, light grey outside, and brilliant yellow inside. Another species, tall Oregon grape (*M. aquifolium*), grows in a more open and dry location, is taller—2 m (6.6 ft.)—and has fewer leaflets (five to nine).

Etymology The species name *aquifolium* means "holly-like."

Habitat Dry coniferous forests in southern coastal BC and Washington.

Season Flowers at the end of May. The berries begin to turn blue by August and persist through autumn.

Traditional Use When steeped, the shredded stems of both species yield a yellow dye that was used in basket making.

Edibility The tart berries are edible and were traditionally mixed with sweeter berries for eating.

▶ OVAL-LEAVED BLUEBERRY

Vaccinium ovalifolium

HEATHER FAMILY Ericaceae

Description The oval-leaved blueberry is one of BC's most recognized and most harvested blueberries. It is a mid-sized bush up to 2 m (6.6 ft.) in height. The bell-shaped pinkish flowers appear before the leaves and are followed by the classic blue berries. Rubbing the berries reveals a covering of dull bloom and a darker berry. The soft green leaves are smooth edged, alternate, and eggshaped (no point) and grow to 4 cm (1.5 in.) in length.

Habitat Moist coniferous forests from sea level to high elevations. Black huckleberry (*V. membranaceum*) and Cascade huckleberry (*V. deliciosum*) also grow in this type of habitat.

Season The fruit ripens as early as July at lower elevations and into September at higher elevations.

Traditional Use As with all blueberries, they were also mashed to create a purple dye used to colour basket materials.

Edibility The berries are edible and were traditionally a valuable and delicious food source. They were eaten fresh, often mixed with other berries, or dried for future use.

▶ RED HUCKLEBERRY

Vaccinium parvifolium

HEATHER FAMILY Ericaceae

Description One of the most graceful of all BC's berry bushes, the red huckleberry grows on old stumps, where it can attain heights of 3–4 m (10–13 ft.). The combination of almost translucent red berries 1 cm (0.5 in.) across, a lacy zigzag branch structure, and oval pale-green leaves 2.5 cm (1 in.) long is unmistakable. The small greenish to pink flowers are inconspicuous.

Habitat Coastal forested areas at lower elevations.

Season Flowering starts mid-April, and the berries ripen by the beginning of July.

Traditional Use The berries' resemblance to salmon eggs made them ideal for fish bait.

Edibility The berries are edible and were traditionally eaten fresh, often mixed with other berries, or dried for winter use.

RED-FLOWERING CURRANT

Ribes sanguineum

CURRANT AND GOOSEBERRY FAMILY Grossulariaceae

Description Red-flowering currant is an upright deciduous bush 1.5–3 m (5–10 ft.) in height. Its flowers range in colour from pale pink to bright crimson and hang in 8–12 cm (3–5 in.) panicles. The bluish-black berries, 1 cm (0.5 in.) across, are inviting to eat but are usually dry and bland. The leaves are 5–10 cm (2–4 in.) across and maple shaped, with three to five lobes. Currants and gooseberries are both in the genus *Ribes*; a distinguishing feature is that currants have no prickles, while gooseberries do.

Habitat Dry open forests at low to mid elevations.

Season Flowers from mid-May to June.

Edibility The berries are edible but very dry tasting.

▶ SALAL

Gaultheria shallon

HEATHER FAMILY Ericaceae

Description Salal is a prostrate to mid-sized bush that grows to 0.5–4 m (1.6–13 ft.) in height. In spring, the small pinkish flowers, 1 cm (0.5 in.) long, hang like strings of tiny Chinese lanterns. The edible dark-purple berries grow to 1 cm (0.5 in.) across and ripen by mid-August or September. Both the flowers and the berries display themselves for several weeks. The dark-green leaves are 7–10 cm (3–4 in.) long, tough, and oval shaped.

Habitat Dry to moist forested areas along the entire coast.

Season Flowering starts at the beginning of May, and the fruit starts to ripen at the beginning of August.

Edibility Though salal is often overlooked by berry pickers, the ripe berries taste excellent fresh and make fine preserves and wine. Traditionally salal was an important food source for most Indigenous peoples. The berries were eaten fresh, often mixed with other berries, or crushed and placed on skunk cabbage leaves to dry. The dried berry cakes were then rolled up and preserved for winter use.

▶ SALMONBERRY

Rubus spectabilis

ROSE FAMILY Rosaceae

Description Salmonberry is one of BC's tallest native berry bushes. Though it averages 2–3 m (6.6–10 ft.) in height, the bush can grow up to 4 m (13 ft.) high. The bell-shaped pink flowers are 4 cm (1.5 in.) across; they bloom at the end of February and are a welcome sight. Flowering continues until June, when both the flowers and the ripe fruit can be seen on the same bush. The soft berries range in colour from yellow to orange to red, with the occasional dark purple, and are shaped like blackberries. The leaves are compound, with three leaflets, much like the leaves of a raspberry. Weak prickles may be seen on the lower portion of the branches; the tops are unarmed.

Etymology The berry's common name comes from its resemblance to the shape and colour of salmon eggs.

Habitat Common on the coast in shaded damp forests.

Season The berries are harvested from mid-June to mid-July.

Edibility The berries are edible. Traditionally, the high water content of the berries prevented them from being stored for any length of time. They were generally eaten shortly after harvesting.

▸ SASKATOON BERRY SERVICE BERRY

Amelanchier alnifolia

ROSE FAMILY Rosaceae

Description Depending on growing conditions, the saskatoon berry can vary from a 1 m (3.3 ft.) shrub to a small tree 7 m (23 ft.) in height. The white showy flowers are 1–3 cm (0.5–1 in.) across and often hang in pendulous clusters. The young reddish berries form early and darken to a purple black by mid-summer. The berries are up to 1 cm (0.5 in.) across. The light bluish-green leaves are deciduous, oval shaped, and toothed above the middle.

Habitat Shorelines, rocky outcrops, and open forests at low to mid elevations.

Season The berries ripen in late June or early July.

Traditional Use The hard, straight wood was a favourite for making arrows.

Edibility The berries are edible and were traditionally eaten fresh, often mixed with other berries, or dried for future use. On the Great Plains, the berries were mashed with buffalo meat to make pemmican.

▶ SOOPOLALLIE SOAPBERRY / CANADIAN BUFFALO BERRY

Shepherdia canadensis

OLEASTER FAMILY Elaeagnaceae

Description Soopolallie is a deciduous bush 1–3 m (3.3–10 ft.) in height. The tiny star-like bronze male and female flowers are borne on separate bushes. The bitter berries are bright red and grow in small clusters along the stems. The leaves and stems are covered with orange dots, giving them a rusty appearance.

Etymology Soopolallie is Chinook for "soapberry," referring to the way the berries froth up when beaten with water.

Habitat Forest edges and upper beaches at low to mid elevations.

Season Flowers from the end of March to April. The berries ripen by mid-July.

Edibility Though edible, the berries are bitter. Among some Indigenous peoples of BC, the berries were beaten—along with other berries for sweetness—in the production of a frothy dish colloquially known as "Indian ice cream."

▶ STINK CURRANT BLUE CURRANT

Ribes bracteosum

CURRANT AND GOOSEBERRY FAMILY Grossulariaceae

Description Stink currant is an unarmed deciduous bush 2–4 m (6.6–13 ft.) in height. The small greenish flowers give way to 15–25 cm (6–10 in.) long clusters of blue-black berries that are covered in a whitish bloom, giving them a blue-grey appearance. The maple-like leaves are up to 20 cm (8 in.) across, with five to seven pointed lobes. When crushed, the leaves emit a musky odour. The name stink currant is not deserved for this attractive plant. Maybe a more fitting name would be musk currant.

Habitat Rich wet soils along creeks and seepage areas at low to subalpine elevations.

Season The berries are ripe by July in lower areas.

Edibility The berries are edible and were traditionally eaten raw or dried into cakes for winter use.

THIMBLEBERRY

Rubus parviflorus

ROSE FAMILY Rosaceae

Description Thimbleberry is an unarmed shrub 3 m (10 ft.) in height. Its large white flowers open up to 5 cm (2 in.) across and are replaced by juicy bright-red berries. The dome-shaped berries are 2 cm (1 in.) across and bear little resemblance to a thimble. The maple-shaped leaves grow up to 25 cm (10 in.) across, and, when needed, make a good substitute for bathroom tissue.

Habitat Common in coastal forests at low to mid elevations.

Season Flowering starts mid-May, and the fruit matures at the end of July to early August.

Traditional Use The large leaves were used to line cooking pits and cover baskets.

Edibility The berries are edible and were traditionally eaten fresh or dried, often mixed with other berries.

TRAILING BLACKBERRY

Rubus ursinus

ROSE FAMILY Rosaceae

Description Trailing blackberry is a moderately armed vine up to 5 m (16 ft.) across. The berries are smaller than the two introduced species, cutleaf and Armenian blackberries, but I find them sweeter, and they ripen earlier.

Habitat Mainly seen at the forest edges, it can tolerate more shade than the two introduced species.

Season Depending on the amount of sunshine, blooming starts at the end of April, and the fruit sets by mid-July.

Edibility Berries are not only edible but delicious, and the leaves can be steeped as a tea. Native to the Pacific Northwest, trailing blackberries have provided nutritious food to the Indigenous people for thousands of years. The berries were eaten by themselves or mixed with other berries for winter food.

WESTERN TEA-BERRY

Gaultheria ovatifolia

HEATHER FAMILY Ericaceae

Description Western tea-berry is a small to prostrate evergreen bush. The bell-shaped flowers are 0.5 cm (0.2 in.) long, white to pinkish, and, like the berries, quite often found hiding beneath the leaves. The red berries are edible, 0.6 cm (0.25 in.) across and grooved into segments. The heart-shaped leaves are alternate, 1.5–4 cm (0.75–1.5 in.) long, finely toothed, shiny above, and dull below. Alpine wintergreen (*G. humifusa*) is an even smaller bush; it is generally seen at higher elevations.

Habitat Moist coniferous forests and bogs at mid to subalpine elevations.

Season Flowers from July to August, depending on elevation.

Edibility The berries are edible. Traditionally, ripe berries were eaten as they were being picked, but they were too sparse to be collected.

WOODLAND STRAWBERRY

Fragaria vesca

ROSE FAMILY Rosaceae

Description The woodland strawberry is an unarmed herbaceous perennial up to 20 cm (8 in.) in height. Its white flowers are 1–3 cm (0.5–1 in.) across and have five petals and a yellow centre. The delicious fruit, 1–3 cm (0.5–1 in.) across, is a smaller version of the cultivated strawberry. The leaves are 3–5 cm (1–2 in.) across and compound, with three coarsely toothed leaflets. The blue-leaf, or wild, strawberry (*F. virginiana*) is similar but has bluish-green leaves, and the terminal teeth on its leaflets are shorter than the teeth on either side; the terminal teeth of the woodland strawberry are longer than the others.

Habitat Found mainly in lower-elevation open forests but can also be found at mid to high elevations.

Season Flowers from May to June.

Edibility The berries are edible, and the leaves can be steeped for tea. Traditionally the juicy berries were eaten fresh rather than dried.

FERNS

BRACKEN FERN

Pteridium aquilinum

POLYPODY FAMILY Polypodiaceae

Description Bracken fern is BC's tallest native fern, often reaching 3 m (10 ft.) or more in height. It is also the most widespread fern in the world. The tall, arching fronds are dark green, with a golden-green stem (stipe). They are triangular and grow singly from rhizomes in spring.

Habitat A diverse growing range, from dry to moist and open to forested regions.

Season Fiddleheads can be picked in April/May while the rhizomes are harvestable anytime.

Edibility The rhizomes can be peeled and eaten fresh or cooked and the fiddleheads boiled and eaten. Bracken has been a traditional food for many cultures around the world and throughout history.

CAUTION Current research suggests that consumption of bracken is not advisable, as it has now been proven to be a potential health hazard and a known carcinogen in some mammals.

LADY FERN

Athyrium filix-femina

POLYPODY FAMILY Polypodiaceae

Description Lady fern is a tall, fragile fern up to 2 m (6.6 ft.) in height. The apple-green fronds average 30 cm (12 in.) across and are widest below the centre, tapering at the top and bottom. The diamond shape distinguishes the lady fern from the similar-looking spiny wood fern (*D. expansa*), whose fronds have an abrupt triangular form. The fronds die off in winter and emerge again in April. The horseshoe-shaped sori appear on the back of the fronds in spring.

Habitat Common in damp forests at low to mid elevations; often associated with deer ferns and spiny wood ferns.

Season Fiddleheads are available in April.

Edibility The young fronds (fiddleheads) are edible and were traditionally steamed and eaten.

LICORICE FERN

Polypodium glycyrrhiza

POLYPODY FAMILY Polypodiaceae

Description Licorice fern is a small evergreen fern commonly seen on mossy slopes and on the trunks of bigleaf maple trees. The dark-green fronds grow to 50 cm (20 in.) long and 5–7 cm (2–3 in.) wide and have a golden stem (stipe). The round spores are produced in a single row under the leaves. The rhizomes have a licorice taste; hence the fern's common name. Leather polypody (*P. scouleri*) is a coastal species that can be seen on the trunks of conifers not far from the ocean. Its appearance is more rounded than that of licorice fern, and its rhizomes are not licorice flavoured.

Habitat Low-elevation forests, where it grows on trunks and branches of large trees, and sometimes on shady outcrops. Commonly seen growing on the trunks of bigleaf maple trees.

Season The roots/rhizomes can be harvested year-round. In dry summers, the leaves will wither and disappear.

Traditional/Medicinal Use The root and rhizomes were traditionally used as a cold and throat medicine.

Edibility The rhizomes can be chewed for flavour and brewed as tea. The roots, too, are edible.

SPINY WOOD FERN

Dryopteris expansa

POLYPODY FAMILY Polypodiaceae

Description Spiny wood fern is an elegant plant up to 1.5 m (5 ft.) tall. The pale-green fronds are triangular, average up to 25 cm (10 in.) across, and die off in winter. In spring, the rounded sori are produced on the underside of the fronds. Spiny wood fern is similar in appearance and requirements (shade, water, and soil conditions) to lady fern (A. *filix-femina*).

Habitat Common in moist forests at low to mid elevations.

Season The roots/rhizomes can be dug up in the autumn.

Edibility The rhizomes are edible. They are baked and then peeled before eating.

WESTERN SWORD FERN

Polystichum munitum

POLYPODY FAMILY Polypodiaceae

Description Western sword fern is the Pacific Northwest's most common fern. It is evergreen and can grow to 1.5 m (5 ft.) in height. The fronds are dark green, with side leaves (pinnae) that are sharply pointed and toothed. A double row of sori forms on the underside of the fronds mid-summer, turning orange by autumn. The fronds are in high demand in eastern Canada and the United States for floral decorations.

Etymology The species name *munitum* means "armed," referring to the side leaves that resemble swords.

Habitat Dry to moist forest near the coast at lower elevations, where it can form pure groves.

Season The roots/rhizomes can be harvested in the spring, while the leaves were collected year-round.

Traditional Use The ferns were used to line steaming pits and baskets and were placed on floors as sleeping mats.

Edibility The larger fleshy roots (rhizomes) can be steamed or baked, then peeled and eaten.

TREES, SHRUBS, AND BUSHES

BALDHIP ROSE WOODLAND ROSE

Rosa gymnocarpa

ROSE FAMILY Rosaceae

Description The baldhip rose is the Pacific Northwest's smallest native rose. It is often prostrate up to 1.5 m (5 ft.) in height. The tiny pink flowers are five petalled, delicately fragrant, 1–2 cm (0.5–1 in.) across, and usually solitary. The compound leaves are smaller than those of the Nootka rose and have five to nine toothed leaflets. The spindly stems are mostly armed, with weak prickles. A good identifier is this rose's unusual habit of losing its sepals, leaving the hip bald.

Etymology The species name *gymnocarpa* means "naked fruit."

Habitat Dry open forests at lower elevations, from southern BC to the redwood forests of California.

Season Flowers in June.

Edibility Rosehips have a higher concentration of vitamin C than oranges and make an excellent jelly or marmalade.

BIGLEAF MAPLE

Acer macrophyllum

MAPLE FAMILY Aceraceae

Description The bigleaf maple is the largest native maple on the West Coast, often exceeding heights of 30 m (100 ft.). Its huge leaves, which are dark green, five lobed, and 20–30 cm (8–12 in.) across, are excellent identifiers. In early spring, it produces beautiful clusters of scented yellow-green flowers, 7–10 cm (3–4 in.) long. The mature winged seeds (samaras), 5 cm (2 in.) long, act as whirligigs when they fall; they are bountiful and an important food source for birds, squirrels, mice, and chipmunks. The fissured brown bark is host to an incredible number of epiphytes, most commonly mosses and licorice ferns.

Habitat Dominant in lower forested areas. Its shallow root system prefers moist soils, mild winters, and cool summers.

Season Flowers from April to May, with the winged seeds seen in July.

Traditional Use The plentiful wood was important in Indigenous culture as a fuel and for carvings, paddles, combs, fish lures, dishes, and handles. The large leaves were used to line berry baskets and steam pits.

Edibility Our Pacific Northwest maples do not have the sugar concentration of the eastern sugar maples. However, in spring our maples produce excellent syrup.

BLACK HAWTHORN

Crataegus douglasii

ROSE FAMILY Rosaceae

Description Black hawthorn is an armed scraggly shrub or bushy tree up to 9 m (30 ft.) in height. The leaves are roughly oval, coarsely toothed above the middle, and 6 cm (2 in.) long. The clusters of white flowers are showy but bland in smell. The edible fruit is purple-black, 1 cm (0.5 in.) long, and hangs in bunches. Older bark is grey, patchy, and very rough.

Habitat Prefers moist soil beside streams, in open forests, or near the ocean.

Season Flowers in May. The berries ripen by mid-June.

Traditional Use The 3 cm (1 in.) thorns were used as tines on herring rakes.

Edibility An excellent jam can be made from black hawthorn berries.

BLUE-BERRIED ELDER BLUE ELDERBERRY

Sambucus caerulea

HONEYSUCKLE FAMILY Caprifoliaceae

Description Blue-berried elder ranges from a bush to a small tree 6 m (20 ft.) in height. Its flowers are similar to those of the red-berried elder but are in flat-topped clusters, not pyramidal. The mature berries are dark blue with a white coating of the bloom, giving them a soft-blue appearance. The leaves are compound, with five to nine oval, lance-shaped leaflets.

Habitat Dry open sites at low elevations. The blue-berried elder is found in inland areas and the Gulf Islands and San Juan Islands.

Season The berries start to turn blue by mid-August and, if not eaten by birds or people, last into October.

Edibility The berries are edible. Though the raw berries are edible, they were traditionally eaten cooked.

CALIFORNIA HAZELNUT BEAKED HAZELNUT

Corylus cornuta var. *californica*

BIRCH FAMILY Betulaceae

Description California hazelnut is a broad spreading shrub 2–5 m (3.3–16 ft.) tall. The male flowers are formed in hanging catkins in early spring. Their pollen is mainly wind distributed to the small female flowers that have beautiful protruding red stigmas. A very close look is needed to see them. The toothed leaves are alternate, 8 cm (3 in.) long, with a heart-shaped base. They give the forest a wonderful autumn-yellow colour.

Habitat Open moist forests along the coast.

Season The hazelnuts are noticeable by the end of June and ripen by August or September.

Edibility The nuts are both edible and delicious. An important food source, they were traditionally traded to areas where the hazelnuts did not grow.

DOUGLAS MAPLE

Acer glabrum

MAPLE FAMILY Aceraceae

Description Douglas maple is a small deciduous tree up to 10 m (33 ft.) in height. Its long palmate leaves are opposite and 5–10 cm (2–4 in.) across and have three to five sharp lobes. The leaves of the closely related vine maple (A. *circinatum*) have seven to nine lobes. The pairs of winged seeds, or samaras, grow to 5 cm (2 in.) across and are joined at right angles.

Habitat Dry open forested sites at low to mid elevations.

Season The leaves turn a brilliant orange in the autumn.

Traditional Use The hard, durable wood was used in many ways, including for snowshoe frames, drum hoops, tongs, throwing sticks, bowls, and masks.

Edibility Our Pacific Northwest maples do not have the sugar concentration of the eastern sugar maples. However, in spring our maples produce excellent syrup.

HAIRY MANZANITA

Arctostaphylos columbiana

HEATHER FAMILY Ericaceae

Description Hairy manzanita is an evergreen shrub 4 m (13 ft.) in height. Its small pinkish-white flowers are 0.7 cm (0.25 in.) long, urn shaped, and formed in terminal clusters. By mid-summer, they develop into mealy brown-red berries. The dull-green leaves are oval shaped, hairy, and up to 5 cm (2 in.) long.

Etymology The common name manzanita is Spanish for "small apples," in reference to the fruit.

Habitat Rocky southwestern-facing areas at low elevations that are exposed to the sun.

Season Flowers in late spring, with berries by mid-summer, between July and August.

Edibility The berries are edible and were traditionally eaten both raw and cooked.

CAUTION The berries are thought to cause serious constipation.

INDIAN PLUM

Oemleria cerasiformis

ROSE FAMILY Rosaceae

Description Indian plum is an upright deciduous shrub or small tree 5 m (16 ft.) in height. Its white flowers, which usually emerge before the leaves, are 1 cm (0.5 in.) across and hang in clusters 6–10 cm (2–4 in.) long. The small plum-like fruit grow to 1 cm (0.5 in.) across; they start off yellowish and red and finish a bluish black. The leaves are broadly lance shaped, light green, and 7–12 cm (3–5 in.) long and appear in upright clusters.

Etymology The species name *cerasiformis* means "cherry shaped," a reference to the fruit.

Habitat Restricted to low elevations on the southern coast and Gulf Islands; prefers moist open broad-leaved forests.

Season Flowers from March to April, with ripe fruit by the end of June.

Edibility They are edible, but a large seed and bitter taste make them better for the birds. Traditionally small amounts were eaten fresh or dried for winter use.

LABRADOR TEA

Rhododendron groenlandicum

HEATHER FAMILY Ericaceae

Description Most of the year, Labrador tea is a small gangly shrub 1.5 m (5 ft.) in height. In spring, the masses of small white flowers turn the shrub into the Cinderella of the bog. The evergreen leaves are lance shaped, alternate, and 4–6 cm (1.5–2 in.) long, with the edges rolled over.

Habitat Peat bogs, lakesides, and permanent wet meadows at low to alpine elevations.

Season Flowering starts mid-May, with the best viewing at the end of May, or early June at lower elevations.

Edibility The leaves can be brewed as a caffeine substitute and have long been used by Indigenous groups across North America as infusions. Caution must be taken—not all people can drink it.

CAUTION Not all people react well to the Labrador tea. Also, it is possible to mistake poisonous bog laurel for Labrador tea; however, the leaves can be distinguished from those of the poisonous bog laurel (*K. microphylla*) by their flat-green colour on top and rust-coloured hairs beneath. To be safe, pick the leaves only when the shrub is in flower.

NOOTKA ROSE

Rosa nutkana

ROSE FAMILY Rosaceae

Description The largest of the Pacific Northwest's native roses, the Nootka rose grows to 3 m (10 ft.) in height. The showy pink flowers are five-petalled, fragrant, 5 cm (2 in.) across, and usually solitary. The compound leaves have five to seven toothed leaflets and are armed with a pair of prickles underneath. The reddish hips are round, plump, and 1–2 cm (0.5–1 in.) across and contrast well with the dark-green foliage.

Habitat Open low-elevation forests throughout BC.

Season Flowering begins mid-May, and the hips start to develop colour by the beginning of August.

Traditional Use Rosehips were strung together to make perfume strings.

Edibility While edible, these rosehips were eaten only in times of famine.

CAUTION The rosehips can cause irritation when passed if their fine fibres are not completely removed before eating.

OCEANSPRAY ARROW-WOOD

Holodiscus discolor

ROSE FAMILY Rosaceae

Description Oceanspray is an upright deciduous shrub 5 m (16 ft.) in height. Its small creamy-white flowers are densely packed to form inverted-pyramidal clusters up to 20 cm (8 in.) long. The fruiting clusters turn an unattractive brown and persist through winter. The leaves are wedge shaped, flat green above, pale green and hairy below, and up to 5 cm (2 in.) long.

Etymology The species name *discolor* refers to the two-coloured leaf.

Habitat Dry open forests at low to mid elevations; often found on rocky outcrops.

Season In full flower by the end of June.

Traditional Use The straight new growth was a favourite for making arrows; hence its common name of arrow-wood. The wood is extremely hard and was used to make harpoon shafts, teepee pins, digging tools, and drum hoops.

Edibility The inconspicuous dried fruits are edible and were cooked by some Indigenous groups. The roots were used to make a mild tea.

PACIFIC CRAB APPLE

Malus fusca

ROSE FAMILY Rosaceae

Description Pacific crab apple is a deciduous shrub or small tree 2–10 m (6.6–33 ft.) in height. Its leaves are 5–10 cm (2–4 in.) long and similar to those of orchard apple trees, except that they often have bottom lobes. The flowers are typical apple blossoms: white to pink, fragrant, and in clusters of five to twelve. The fruit that follows is 1–2 cm (0.5–1 in.) across and green at first, turning yellowy reddish. On older trees, the bark is scaly and deeply fissured. The Pacific crab apple is the West Coast's only native apple.

Habitat High beaches, moist open forests, swamps, and stream banks at lower elevations.

Season Flowers from the end of April to May.

Traditional Use The hard wood was used to make digging sticks, bows, handles, and halibut hooks.

Edibility The small apples are edible and were an important traditional food source.

PACIFIC SILVER FIR

Abies amabilis

PINE FAMILY Pinaceae

Description Pacific silver fir is a straight-trunked, symmetrical conifer up to 60 m (200 ft.) in height. The bark on young trees is smooth grey, with prominent resin blisters. As the tree ages, the bark becomes scaly, rougher, and often lighter in colour. The cones are dark purple, barrel shaped, and up to 12 cm (5 in.) long. They sit erect on the upper portion of the tree. The needles are a lustrous dark green on the upper surface and silvery white below, with a notched tip.

Etymology The species name *amabilis* means "lovely fir."

Habitat Moist forests at mid to high elevations.

Season Sap collection is usually done in early spring.

Traditional Use The soft wood was used for fuel, but little else.

Edibility The sap of the Pacific silver fir can be chewed like gum.

PAPER BIRCH CANOE BIRCH

Betula papyrifera

BIRCH FAMILY Betulaceae

Description Paper birch is a medium-sized tree reaching heights of 20 m (66 ft.). Its serrated leaves are 8–12 cm (3–5 in.) long, rounded at the bottom, and sharply pointed at the apex. Male and female catkins can be seen in early spring, just before the leaves appear. The white peeling bark is a good identifier on younger trees. There is also a red-bark variety that can be confused with the native bitter cherry (*P. emarginata*).

Etymology The species name *papyrifera* means "to bear paper."

Habitat Rare in low-elevation coastal forests but common in interior forests; prefers moist soil and will tolerate wet sites.

Season The best time to collect the sap is at the end of winter or early spring.

Traditional Use The bark was used to make canoes, cradles, food containers, writing paper, and coverings for teepees. The straight-grained wood was used for arrows, spears, snowshoes, sleds, and masks.

Edibility The sap when collected in spring makes an excellent syrup.

RED ALDER

Alnus rubra

BIRCH FAMILY Betulaceae

Description The largest native alder in North America, the red alder grows quickly and can reach 25 m (81 ft.) in height. Its leaves are oval shaped, grass green, and 7–15 cm (3–6 in.) long, with a coarsely serrated edge. Hanging male catkins, 7–15 cm (3–6 in.) long, decorate the bare trees in early spring. The fruit (cones) are 1.5–2.5 cm (0.75–1 in.) long; they start off green, then turn brown, and persist through winter. The bark is thin and grey on younger trees and scaly when older. Red alder leaves give a poor colour display in autumn, when they are mainly green or brown.

Habitat Moist wooded areas, disturbed sites, and stream banks at low to mid elevations.

Season The fresh buds and catkins are available in early spring.

Traditional Use The soft straight-grained wood is easily worked and was used for making masks, bowls, rattles, paddles, and spoons. The red bark was used to dye fishnets, buckskins, and basket material.

Edibility The fresh buds and young catkins (male flowers) are not considered a delicacy but are good for emergencies.

RED-BERRIED ELDER RED ELDERBERRY

Sambucus racemosa

HONEYSUCKLE FAMILY Caprifoliaceae

Description Red-berried elder is a bushy shrub up to 6 m (20 ft.) in height. Its small flowers are creamy white and grow in pyramidal clusters 10–20 cm (4–8 in.) long. The berries that replace them take up to three months to turn bright red and are a favourite food for birds. The leaves are compound and 5–15 cm (2–6 in.) long, with five to nine opposite leaves.

Habitat Moist coastal forest edges and roadsides.

Season Flowers in May and sets mature berries in July.

Traditional Use The pithy branches were hollowed out and used as blowguns.

Edibility The berries are edible when either steamed or boiled.

CAUTION The berries are poisonous to humans when eaten raw.

REDSTEM CEANOTHUS

Ceanothus sanguineus

BUCKTHORN FAMILY Rhamnaceae

Description Redstem ceanothus is a deciduous bush up to 3 m (10 ft.) in height. Its fragrant tiny flowers grow in terminal clusters 10–15 cm (4–6 in.) long. The oval leaves are finely toothed, up to 10 cm (4 in.) long, with three major veins.

Etymology The species name *sanguineus* means "bloody red," referring to the colour of the new twigs.

Habitat Open forests and forest edges on dry sites, at low to mid elevations. Snowbush (*C. velutinus*) also grows in these habitats and blooms at about the same time.

Season Blooms from the end of May to June.

Traditional Use The wood was sometimes used in smoking deer meat.

Edibility A well-known tea substitute can be made from the dried leaves and flowers.

ROWAN WITCH WOOD / EUROPEAN MOUNTAIN ASH

Sorbus aucuparia

ROSE FAMILY Rosaceae

Syn. *Pyrus aucuparia*

Description Rowan is a small to medium-sized tree that rarely exceeds 13 m (43 ft.) in height. Its flowers are creamy white, have five petals, and are borne in flat-topped clusters. The berries are 1 cm (0.5 in.) across, orange red, and, upon close inspection, shaped like miniature apples. The leaves are alternating, finely sawtoothed, and pinnately compounded into eleven to fifteen leaflets.

Origin A European garden escapee, this introduced species has naturalized well, and its berries are a favourite with birds.

Etymology The species name *aucuparia* is from the Latin word *aucupor* "bird catching" as the berries were used as bait to catch birds. For centuries, rowans were planted near English houses to protect the families from witches; hence the common name witch wood.

Habitat Rowan can be found at lower elevations, especially near townships.

Season Flowers from April to May, with the berries ripening from September to October.

Traditional Use Rowan is rich in history. In the United Kingdom, it was planted near homes to protect owners from witches and in cemeteries to keep the dead in their graves. Christ is believed to have been crucified on a cross made from mountain ash, cedar, holly pine, or cypress.

Edibility The berries are edible but bitter. Rowan berries have been used in Scotland for hundreds of years for making Rowan jelly.

SHORE PINE

Pinus contorta

PINE FAMILY Pinaceae

Description Depending on where they are growing, shore pines vary dramatically in size and shape. By the shoreline, they are usually stunted and twisted from harsh winds and nutrient-deficient soil. A little farther inland, they can be straight trunked and up to 20 m (66 ft.) in height. The small cones, 3–5 cm (1–2 in.) long, are often slightly lopsided and remain on the tree for many years. The dark-green needles are 4–7 cm (2–3 in.) long and grow in bundles of two. Another variety, lodgepole pine (*P. contorta* var. *latifolia*), grows straighter and taller, up to 40 m (132 ft.)

Habitat The coastal variety grows in the driest and wettest sites, from low to high elevations.

Season The nuts of the shore pine can be collected year-round. In terms of definition, a pine produces cones; the cones contain nuts with a hard shell; within the nut is the edible seed.

Traditional Use The straight wood was used for teepee poles, torches, and arrow and harpoon shafts.

Edibility The nuts are edible but small and hard to reach.

SITKA ALDER

Alnus sinuata

BIRCH FAMILY Betulaceae

Description The Sitka alder is a deciduous shrub or small tree 3–7 m (10–23 ft.) high. Its coarse leaves are double serrated, grass green, and 7–10 cm (3–4 in.) long. In early spring, it becomes covered in pollen-producing catkins 10–15 cm (4–6 in.) long and female cones 2 cm (1 in.) long.

Habitat As with most alders, it prefers moist conditions, from the coast of the Arctic Circle to the high mountains of California. Often grows on avalanche sites.

Season In leaf from April to November.

Traditional Use The soft straight-grained wood is easily worked and was used for making masks, bowls, rattles, paddles, and spoons. The red bark was used to dye fishnets, buckskins, and basket material.

Edibility The fresh buds and young catkins (male flowers) are not considered a delicacy but are good for emergencies.

SITKA MOUNTAIN ASH

Sorbus sitchensis

ROSE FAMILY Rosaceae

Description Sitka mountain ash is a small multi-stemmed bush or thicket 1.5–4.5 m (5–15 ft.) in height. Its compound bluish-green leaves have seven to thirteen leaflets, eleven being the norm. The tiny white flowers are in terminal clusters, 5–10 cm (2–4 in.) across. From August to September, the bushes and trees display a wonderful show of bright red-orange berries. The bark is thin and shiny grey. The native mountain ash should not be confused with the larger European mountain ash or rowan (*S. aucuparia*).

Habitat Sitka mountain ash stays primarily where its name suggests: in the mountains.

Season Flowers in the spring, followed by masses of orange berries by the end of summer.

Edibility The berries are edible either raw or cooked.

SITKA SPRUCE

Picea sitchensis

PINE FAMILY Pinaceae

Description Sitka spruce is often seen on rocky outcrops as a twisted dwarf tree, though in favourable conditions it can exceed 90 m (300 ft.) in height. Its reddish-brown bark is thin and patchy, a good identifier when the branches are too high to observe. The cones are golden brown and up to 8 cm (3 in.) long. The needles are dark green, up to 3 cm (1 in.) long, and sharp to touch. Sitka spruce has the highest strength-to-weight ratio of any BC, Washington, or Oregon tree. It was used to build the frame of Howard Hughes's infamous plane, the *Spruce Goose*.

Habitat A temperate rainforest tree that does not grow farther than 200 km (125 mi.) from the ocean.

Season The best time to collect the sap is in the early spring. New shoots are usually visible by May.

Traditional/Medicinal Use The new shoots and inner bark were a good source of vitamin C. Also, the best baskets and hats were woven from spruce roots.

Edibility The pitch (sap) can be chewed as a gum.

SWAMP ROSE CLUSTERED WILD ROSE

Rosa pisocarpa

ROSE FAMILY Rosaceae

Description Swamp rose is a weakly armed bush up to 1.5 m (5 ft.) tall. Its pale-pink flowers are formed in small clusters (not singly); hence the common name of clustered wild rose. The hips are small and pea sized. The leaves consist of five to seven pointed leaflets with two thorns at the base.

Etymology The species name *pisocarpa* means "with pea-like fruit."

Habitat Low-lying swampy areas along the coast.

Season Flowers from May to June.

Edibility The petals can be steeped in hot water with a little honey, then cooled down for a refreshing afternoon drink.

WESTERN WHITE PINE

Pinus monticola

PINE FAMILY Pinaceae

Description Western white pine is a medium-sized symmetrical conifer up to 40 m (132 ft.) in height. Its bark is silvery green grey when young and dark brown and scaly when old. The cones are 15–25 cm (6–10 in.) long and slightly curved. The bluish-green needles are 5–10 cm (2–4 in.) long and grow in bundles of five.

Etymology The species name *monticola* means "growing on mountains."

Habitat Moist to wet soils on the southern coast at low to high elevations.

Season The best season for collecting the bark and sap is the first warm day in spring. The sap starts rising and the bark becomes slippery.

Traditional Use The bark was peeled into strips and sewn together with roots to make canoes. The pitch was used for waterproofing.

Edibility As with all our Pacific Northwest pines, the seeds are edible. The sap resin can be chewed like a gum.

WHITEBARK PINE

Pinus albicaulis

PINE FAMILY Pinaceae

Description Whitebark pine can reach heights of 20 m (66 ft.), but it is often seen as a stunted bush under 3 m (10 ft.) Its needles are slightly curved, up to 8 cm (3 in.) long, purplish when young, and green when mature.

Habitat Exposed dry sites at subalpine elevations. A related species, western white pine (*P. monticola*) can be seen at slightly lower elevations.

Season The seeds are available year-round.

Edibility The seeds, though small and painfully hard to get at, are edible.

WHITE-FLOWERED RHODODENDRON

Rhododendron albiflorum

HEATHER FAMILY Ericaceae

Description White-flowered rhododendron is a deciduous bush up to 2 m (6.6 ft.) in height. Its showy white-cream flowers are 2 cm (1 in.) long and grow in clusters of two to four along the stems. The leaves are alternate and 3–7 cm (1–2 in.) long, and the upper surface is shiny but also slightly hairy.

Habitat Moist open slopes or on the edges of coniferous forests at subalpine elevations.

Season Flowers from mid-July to the beginning of August.

Traditional/Medicinal Use The buds were boiled by some Indigenous groups to be used as a sore-throat remedy or chewed to treat stomach ulcers.

Edibility The leaves can used for tea.

FUNGI AND ALLIES

▶ ADMIRABLE BOLETE

Boletus mirabilis

BOLETE FAMILY Boletaceae

Description Admirable bolete is probably the best looking of the bolete species. The cap is up to 15 cm (6 in.) across and wine red to dark red brown, with a velvet-like surface. The underside tubes are initially pale yellow and mature to a darker greenish yellow. The stalk is up to 17 cm (7 in.) in height, reddish brown, and generally wider at the base.

Etymology The species name *mirabilis* means "wonderful" or "remarkable," which it is.

Habitat Coniferous forests; usually associated with decaying hemlock.

Season Fruitbodies are seen from late summer to autumn.

Edibility Edible and sought after.

▶ ALCOHOL INKY TIPPLER'S BANE

Coprinus atramentarius

INK CAP FAMILY Coprinaceae

Description Alcohol inky is a very common but short-lived mushroom. The elongated bell-shaped cap is up to 10 cm (4 in.) across and pale grey with brownish tints. The stalk and gills are white to pinkish; the stalk is up to 15 cm (6 in.) in height. The entire fruitbody turns black as it ages.

Etymology The species name *atramentarius* means "inky."

Habitat Forested areas, lawns, roadsides, and old stumps.

Season Fruitbodies can be seen from May to November.

Traditional Use An ink was made from the spent caps.

Edibility This mushroom is edible and mild tasting.

CAUTION This mushroom should not be eaten by people who consume alcohol. Even a tiny portion of alcohol before, during, or after eating the mushroom will set off poisonous reactions. The common name of Tippler's bane suggests this.

▶ CAULIFLOWER MUSHROOM

Sparassis crispa

CORAL AND CLUB FUNGI FAMILY Clavariaceae

Description Cauliflower mushroom is a wood-rotting fungus that is mainly associated with Douglas fir and pine in the Pacific Northwest. The fruitbodies can become very large, up to 50 cm (20 in.) across by 50 cm (20 in.) tall, and weigh up to 20 kg (45 lb). The fruitbodies are densely branched with wavy flesh-like leaves; they are white to cream coloured, becoming light brown with age.

Etymology The species name *crispa* means "finely waved" or "closely curled."

Habitat In coniferous forests on or near stumps or at the base of living trees.

Season Fruitbodies are seen from August to the end of October.

Edibility It is sought after when fresh.

CRESTED CORAL FUNGUS

Clavulina coralloides

CORAL AND CLUB FUNGI FAMILY Clavariaceae

Description Crested coral fungus is very widespread across North America and Europe. The fruitbodies are white when young, purplish brown when mature, many forked or crested, and up to 10 cm (4 in.) across and in height.

Habitat On the ground or on extremely rotted conifer wood.

Season Fruitbodies are seen from July to late October.

Edibility Considered edible only in its pure-white stage.

▶ FRIED CHICKEN MUSHROOM

Lyophyllum decastes

DIVERSE FUNGI FAMILY Tricholomataceae

Description Fried chicken mushrooms grow in dense clusters on the ground. The caps are 5–12 cm (2–5 in.) across and range from yellowish brown to greyish brown to dark brown, depending on age. The gills start out white and then turn yellowish when mature. The stalks are 5–10 cm (2–4 in.) in height, whitish, and darker toward the base.

Habitat On the ground in open areas, at forest edges, and on trails that are covered in bark mulch.

Season Fruitbodies are seen from mid-June to the end of October.

Edibility An edible mushroom with a mild, pleasant taste.

CAUTION A lot of people search out this mushroom and have no problems. However, there are reports of others having gastric upsets. Caution should be taken.

▶ GEM-STUDDED PUFFBALL COMMON PUFFBALL

Lycoperdon perlatum

PUFFBALL FAMILY Lycoperdaceae

Description Gem-studded puffball is probably the Pacific Northwest's most common puffball. The rounded fruitbodies are up to 8 cm (3 in.) across and 10 cm (4 in.) in height, white to start, and light brown at maturity. Their surface is covered with different-sized conical spines that eventually fall off, leaving a net-like pattern. As with most puffballs, a hole develops on the top for the spores to puff out.

Etymology The genus name *Lycoperdon* means "wolf fart"; I don't know who thought this one up.

Habitat Common on the ground in open forests and on roadsides.

Season Fruitbodies are seen from August to November.

Edibility Edible when young and pure white inside.

CAUTION In its button stage, the highly poisonous death cap mushroom (A. *phalloided*) can be mistaken for a puffball. When collecting puffballs, cut them in half—the insides should be all white, with no gills.

▶ GOLDEN CHANTERELLE

Cantharellus formosus

CHANTERELLE FAMILY Cantharellaceae

Description Golden chanterelle is probably the most sought-after edible fungus in the Pacific Northwest. The golden caps are initially convex and then become depressed or funnel shaped, up to 13 cm (5 in.) across. The undersides have gill-like ridges that run down to the stem. The stem is up to 10 cm (4 in.) in height. Both the stem and the undersides are the same colour as the cap. Golden chanterelle is the state mushroom of Oregon.

Etymology The species name *formosus* means "beautiful."

Habitat Coniferous forests.

Season Fruitbodies are seen from July to November.

Edibility Edible and sought after. In the Pacific Northwest, golden chanterelles are at their peak in October, when we harvest them to make our delicious Thanksgiving stuffing.

LION'S MANE MUSHROOM

Hericium erinaceus

TOOTH FUNGI FAMILY Hydnaceae

Description It's not hard to see why the common name lion's mane was given to this fungus, especially when it is yellowish gold. The fruitbodies are up to 40 cm (16 in.) across and 20 cm (8 in.) long. The spines are up to 8 cm (3 in.) long; they start off white, turn to a yellowish gold, and then finish a mud brown.

Etymology The species name *erinaceus* means "resembling a hedgehog." Bearded hedgehog mushroom is another common name for this fungus.

Habitat In hardwood forests, mainly on maples and oaks. I have found lion's mane on older living beech trees in parks.

Season Fruitbodies are seen from late August to the beginning of November.

Edibility Edible only when bright white.

▶ ORANGE PEEL FUNGUS

Aleuria aurantia

CUP FUNGI FAMILY Pyronemataceae

Description When seen in grassy parks, orange peel fungus does look like orange peels discarded by picnickers. The rubbery fruitbodies are up to 10 cm (4 in.) across by 5 cm (2 in.) in height. The inner cap is smooth and bright orange, while the outside is covered in fine white powder, making it appear lighter in colour.

Habitat Open areas, lawns, parks, and the sides of gravelly paths.

Season Fruitbodies are seen in late spring and autumn.

Edibility Edible; however, it is said to be tasteless.

▸ PEAR-SHAPED PUFFBALL

Lycoperdon pyriforme

PUFFBALL FAMILY Lycoperdaceae

Description Pear-shaped puffballs produce white- to cream-coloured fruitbodies that mature to light brown. They are up to 4 cm (1.5 in.) across and 6 cm (2.5 in.) in height; they have a slightly granular surface when young and are smooth when mature. A hole develops on the top for the spores to puff out.

Etymology The species name *pyriforme* means "pear shaped."

Habitat Mixed forests, on decayed stumps and buried wood.

Season Fruitbodies are seen from July to early November.

Edibility Edible when young and pure white inside.

CAUTION In its button stage, the highly poisonous death cap mushroom (A. *phalloided*) can be mistaken for a puffball. When collecting puffballs, cut them in half—the insides should be all white, with no gills.

PINE MUSHROOM

Tricholoma murrillianum

DIVERSE FUNGI FAMILY Tricholomataceae

Description When young, pine mushrooms are white, but they soon start turning brown. The gilled caps are up to 5–10 cm (2–4 in.) across. The overall height (cap and stem) can be up to 12 cm (5 in.).

Habitat Usually seen in coniferous forests where it is thought the pine mushrooms have a symbiotic relationship with the conifers and arbutus.

Season Fruitbodies are mainly found in September and October, when the temperature lowers and moisture returns.

Edibility Very choice and sought after by mushroom aficionados.

▸ PRINCE

Agaricus augustus

MEADOW MUSHROOM FAMILY Agaricaceae

Description The prince mushroom is so called because of its size and because, when cooked, it makes a meal fit for a prince. The enormous cap, up to 30 cm (12 in.) across, is initially rounded, becomes convex, and then flattens out completely. The surface is cream coloured with large reddish-brown scales that increase in size toward the centre. The gills start off pinkish grey and then turn dark brown at maturity. The stalk is up to 20 cm (8 in.) in height and has a large pendulous ring; the stalk is smooth above the ring, with scales below.

Habitat Open coniferous forests and parks, especially under true cedars.

Season Fruitbodies are seen from July to November.

Edibility Choice and well sought after. The flesh has an almond aroma.

▶ SHAGGY MANE SHAGGY INK CAP

Coprinus comatus

SHAGGY CAP FAMILY Coprinaceae

Description Shaggy mane is another aptly named edible mushroom. The cylindrical shaggy cap is white with a brown top and up to 8 cm (3 in.) across. It is supported on a smooth white stalk up to 20 cm (8 in.) in height. As the mushroom ages, the white cap and gills start turning inky black from the bottom up; hence the common name of the shaggy ink cap.

Habitat Open mixed forests, lawns, and on hard-packed soils on roadsides and trailsides. The best concentrations are seen when the rain arrives in late summer and autumn.

Season Shaggy mane can be seen from April to June and from September to November.

Edibility Considered one of the best eating mushrooms when young and white. When the caps start showing their inky transition, they are too late to be edible.

▶ SMOKY GILLED WOODLOVER

Hypholoma capnoides

STROPHARIA FAMILY Strophariaceae

Description Smoky gilled woodlover, as its name indicates, grows on decaying wood. The caps vary in colour from yellow orange to brownish and grow up to 6 cm (2.5 in.) across. The gills are smoky grey when young and darker purple brown when mature. The stalk is up to 10 cm (4 in.) in height and yellowish toward the cap, darkening to brownish toward the base.

Habitat Mixed forests on conifer stumps and logs. In ideal conditions, there can be hundreds in the same location.

Season Fruitbodies are seen from mid-August to the beginning of December.

Edibility Edible and sought after.

CAUTION Care should be taken not to confuse it with the poisonous species sulphur tuft (*H. fasciculare*), the gills of which age from yellow to greenish.

SULPHUR SHELF

Laetiporus conifericola

POLYPORE FAMILY Polyporaceae

Description Sulphur shelf is one of the largest and most beautiful fungi in the Pacific Northwest. There are two species in the Pacific Northwest: *L. conifericola*, which grows on conifers, and *L. gilbertsonii*, which grows on hardwoods. The tops of the fan-shaped shelves are orange, and the undersides are sulphur yellow to orange. The shelves can independently grow to 60 cm (24 in.) across and can collectively be several metres or feet across. The shelves are rubbery when young but fade to white yellow and get crumbly when old. Sulphur shelves are annuals.

Habitat Both species grow in mixed forests; the fruitbodies are usually seen in dead wood. I came across a beautiful patch of *L. conifericola* on an amabilis fir log at an elevation of 1,200 m (3,900 ft.).

Season I have found the fresh fruitbodies only from September to November.

Edibility Young specimens or the fresh margin or older specimens are the most edible. Best eaten cooked.

▶ VANCOUVER GROUNDCONE

Boschniakia hookeri

BROOMRAPE FAMILY Orobanchaceae

Description As opposed to dwarf mistletoe, which is parasitic on the above-ground branches of conifers, Vancouver groundcone is a parasite on the roots of salal. It is an herbaceous perennial up to 13 cm (5 in.) tall and has a mix of colours, including yellow, orange, and purple. The small tube-shaped flowers are yellow to purple and protrude above the scale-like leaves. The overall size and shape of this plant does look like a malformed cone that has dropped to the ground.

Habitat Low elevations. You will most likely see salal growing above or nearby.

Season Flowers in the spring.

Edibility The potato-like stem base can be eaten raw.

▶ WHITE FAIRY FINGERS

Clavaria vermicularis

CORAL AND CLUB FUNGI FAMILY Clavariaceae

Description White fairy fingers are also known as white worm coral and white spindles. The white fruitbodies are slender, fragile, unbranched, and up to 15 cm (6 in.) in height. Purple fairy clubs are very similar except for their colour, and they are shorter.

Etymology The species name *vermicularis* means "worm-like."

Habitat On the ground in open conifer forests. I have found purple fairy clubs growing in elevations of up to 1,500 m (4,950 ft.) in pure fir stands.

Season Fruitbodies are seen from July to September, depending on elevation.

Edibility Both species are considered edible when young.

▸ YELLOW MOREL MUSHROOM

Morchella esculenta

MOREL FAMILY Morchellaceae

Description The yellow morel is a fruiting mushroom up to 2–10 cm (1–4 in.) tall. Depending on its age and growing medium, the caps can range from yellow to cream coloured to greyish brown. The caps are undulating and sponge-like, hollow and brittle when cut or broken.

Etymology The species name *esculenta* means "edible."

Habitat Forest fire affected areas and open forests with no tall growth.

Season Usually mid-May through June, depending on how far north you are.

Edibility Morels appear atop the list for mushroom pickers and consumers. The mushrooms should be pan fried or roasted before eating and any decay noticed should be removed. Again, try eating just a small amount of the fruiting body to see if it agrees with you.

ZELLER'S BOLETE

Boletus zelleri

BOLETE FAMILY Boletaceae

Description Zeller's bolete is one of the more common boletes in the Pacific Northwest. The cap is 13 cm (5 in.) across and dark reddish brown to almost black, with a wrinkled, velvety surface. The edge of the cap is often cream coloured. The tubes on the underside are yellowish. The stalk is 11 cm (4.5 in.) in height and red or yellowish with red lines.

Etymology The species name *zelleri* commemorates Professor Sanford Myron Zeller (1885–1948), who first discovered it in Seattle, Washington.

Habitat Coniferous forests and forest edges.

Season Fruitbodies are seen from summer to autumn.

Edibility Edible, with a mild taste.

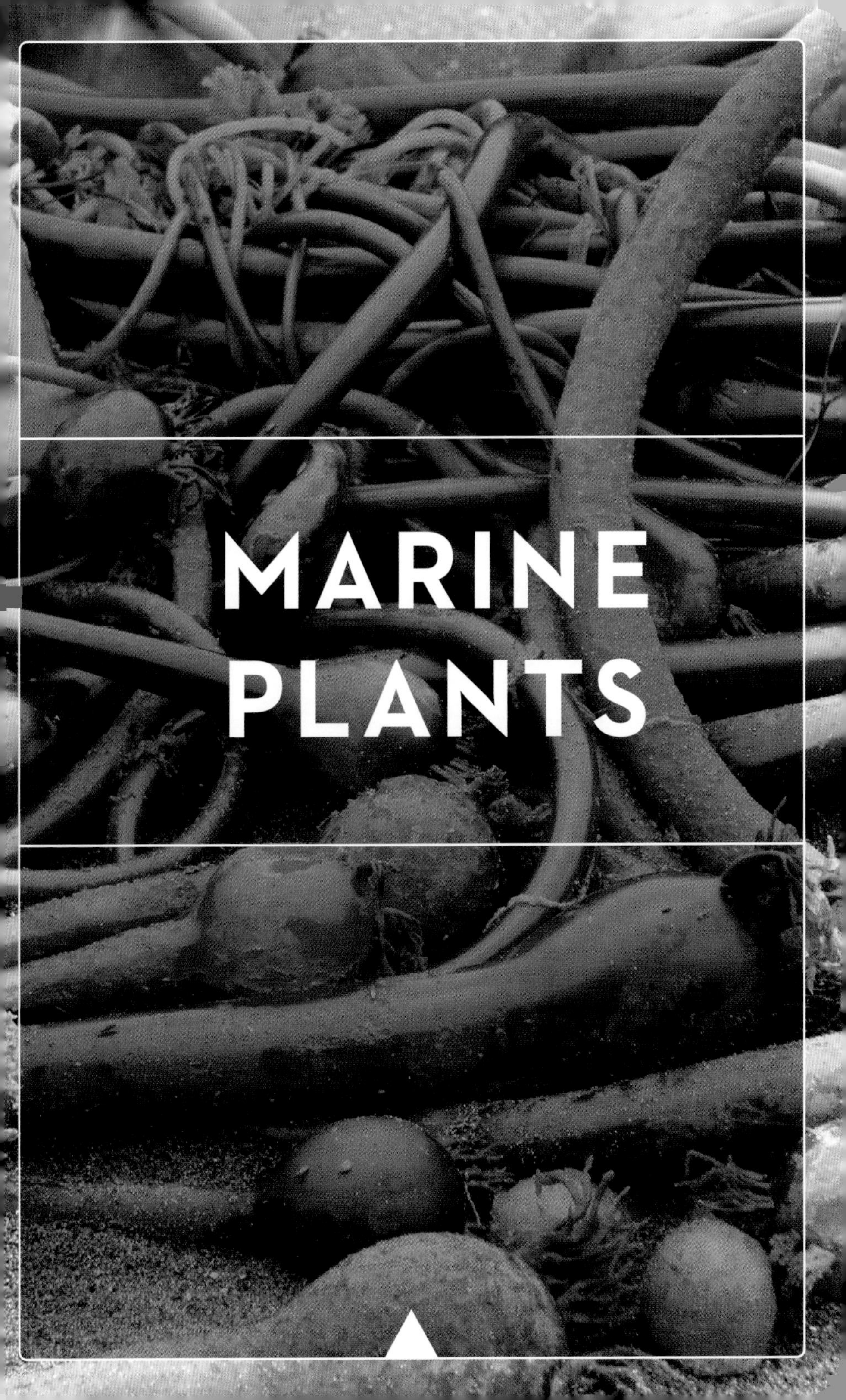

MARINE PLANTS

▶ BULL KELP

Nereocystis luetkeana

BROWN KELP FAMILY Laminariaceae

Description Bull kelp is one of the most recognized kelps in the Pacific Northwest and one of the largest, attaining lengths of over 30 m (100 ft.). The long brown stipe is kept afloat by a large pneumatocyst (float), which is decorated with over twenty blades, each up to 3 m (10 ft.).

Etymology The genus name *Nereocystis* is Greek for "mermaid's bladder."

Habitat Grows on rocky areas in the lower intertidal zone, from Alaska to California.

Season As long as they are floating, you can collect and clean the stipes and bulbs from spring to autumn.

Edibility The stipes and bulb make very tasty pickles, and the blades can be dried for snacks and later use or used fresh in soups and stews.

▶ MARINE EELGRASS

Zostera marina

EELGRASS FAMILY Zosteraceae

Description Marine eelgrass has leaves up to 1.2 m (4 ft.) long and 1 cm (0.5 in.) wide. A dwarf introduced eelgrass (Z. *japonica*) can often be seen growing with common eelgrass. The dwarf eelgrass has shorter, thinner leaves.

Habitat Prefers protected bays and mud flats. Found from Alaska to Mexico.

Season Throughout the summer.

Edibility The seeds and rhizomes are edible and were harvested by West Coast Indigenous peoples.

▶ SCOULER'S SURF GRASS

Phyllospadix scouleri

EELGRASS FAMILY Zosteraceae

Description Scouler's surf grass leaves are up to 1.5 m (5 ft.) long and only 0.4 cm (0.15 in.) wide. Male and female flowers are borne on separate plants (dioecious).

Habitat Grows on rocky shores in very exposed areas of the lower intertidal zone from Alaska to Mexico.

Season Can be harvested throughout the summer.

Edibility The rhizomes are edible and were harvested by West Coast Indigenous peoples.

▶ SEA ASPARAGUS AMERICAN GLASSWORT

Salicornia virginica

GOOSEFOOT FAMILY Chenopodiaceae

Description Sea asparagus is an edible perennial up to 30 cm (12 in.) in height. Its tiny yellow-green flowers grow in threes in small sunken cavities along the succulent stems. The scale-like leaves are almost non-existent.

Habitat Common along the BC and Washington coastline in areas with little wave action. It can form dense colonies of up to hundreds of square metres or feet in tidal flats and salt marshes.

Season Flowers from July to August.

Edibility As its name suggests, the young stems of this wild green vegetable can be collected and eaten raw or cooked. They have a salty flavour, but it can be masked with a few herbs.

▶ SEA HAIR

Enteromorpha sp.

GREEN SEAWEED FAMILY Ulvaceae

Description Sea hair has hollow yellowish-green blades up to 20 cm (8 in.) long. The blades are so thin they are often referred to as maiden hair.

Etymology The genus name *Enteromorpha* is Greek for "intestine form."

Habitat Can be found on rocky areas in the mid to upper intertidal zone, from Alaska to Mexico.

Season So long as it is still green, it is harvestable.

Edibility I have used small amounts in my soups, stews, and spaghetti sauce.

▶ SEA LETTUCE

Ulva fenestrata

GREEN SEAWEED FAMILY Ulvaceae

Description Sea lettuce is bright green and to a length of 51 cm (20 in.). The blades are so delicate that they are translucent when looked at in the sunlight.

Etymology The genus name *Ulva* is Latin for "marsh plant."

Habitat Tide pools and rocky shorelines in the lower to upper intertidal zones, from Alaska to California.

Season Mid-summer.

Edibility I have washed and dried sea lettuce and used it as a snack.

▶ SEA PALM

Postelsia palmaeformis

FAMILY Laminariaceae

Description Sea palm is an apt name for this aquatic palm tree. It is light green when young and brown when mature, and it grows to 60 cm (24 in.) in height. The serrated blades hang over the top of the stipe, giving it a palm tree–like appearance.

Etymology The genus name *Postelsia* commemorates Alexander Filippovich Postels (1801–71), an Estonian naturalist hired by Czar Nicholas I of Russia to explore the Pacific coast.

Habitat Grows on wave-swept rocky shores in the mid to lower intertidal zone, from BC to California.

Season Can be seen on ocean ledges just offshore from summer to autumn.

Edibility Harvesting sea palm is now illegal; however, after some severe summer storms, I have gathered some that has come ashore. The blades that I cleaned and ate raw or dried tasted very much like bull kelp.

▶ SEERSUCKER RIBBED KELP

Costaria costata

RIBBED KELP FAMILY Costariaceae

Description Seersucker is light brown to dark brown in colour, with elliptical blades up to 2 m (6.6 ft.) long and five defined ribs. When growing in the semi-protected waters of the lower intertidal zone, the blades are wider than when growing in harsher waters.

Etymology The genus name *Costaria* and species name *costata* are both Latin for "rib."

Habitat Found in the lower intertidal zone, from Alaska to California.

Season May and June are usually the best time to harvest before they become too big and tough.

Edibility Edible when young. I have eaten the thin pieces between the stipes and have lived for another day.

▶ SMALL PERENNIAL KELP GIANT KELP

Macrocystis integrifolia

BROWN KELP FAMILY Laminariaceae

Description Small perennial kelp is yellowish to dark brown in colour and can grow to 30 m (100 ft.) in length. The individual blades are up to 35 cm (14 in.) long and have a pointed float at the end.

Etymology The genus name *Macrocystis* is Greek for "giant bladder."

Habitat Grows in rocky areas in the lower intertidal zone, from Alaska to California.

Season Can be found washed ashore mid-summer to autumn.

Edibility I find this kelp is best dried and eaten. Some people get stomach aches when they eat this kelp. I think, in the raw form, it is hard for the stomach to break it down.

▶ SPLIT KELP

Laminaria setchellii

BROWN KELP FAMILY Laminariaceae

Description Split kelp is dark brown to almost black in colour. The stipe can grow to 80 cm (31 in.) high with terminal split blades up to 80 cm (31 in.) long.

Etymology The genus name *Laminaria* is Latin for "thin leaf."

Habitat Prefers wave-swept exposed rocky areas in the lower intertidal zone, from Alaska to California.

Season Can be found washed ashore mid-summer to autumn.

Edibility The blades can be dried or boiled and used in stews and sauces.

▶ SUGAR WRACK

Laminaria saccharina

BROWN KELP FAMILY Laminariaceae

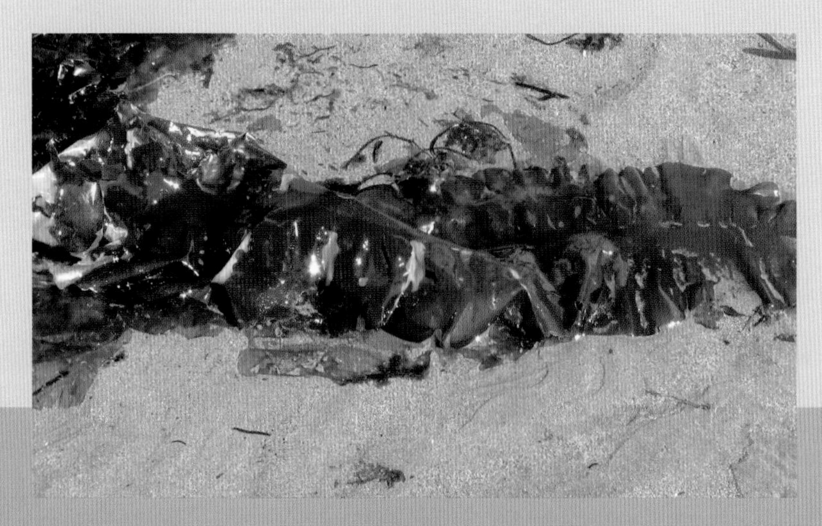

Description Sugar wrack is so called because of its sugary taste. The blades are dark brown and grow to 3 m (10 ft.) in length. They are held firmly in place by a small stipe and holdfast.

Etymology The species name *saccharina* is Latin for "sugar."

Habitat Found in rocky areas in the lower intertidal zone, from Alaska to California.

Season Can be found washed ashore mid-summer to autumn.

Edibility Like split kelp, the blades can be dried or boiled and used to flavour stews and sauces, giving a sweeter taste than other kelps.

▶ WINGED KELP

Alaria marginata

BROWN KELP FAMILY Alariaceae

Description Winged kelp is so named for the spore-producing winged blades growing from its base. The main blade has a midrib and is light brown to dark brown, wavy edged, and up to 3 m (10 ft.) in length.

Etymology The genus name *Alaria* is Latin for "wine."

Habitat Rocky areas in the lower intertidal zone, from Alaska to California.

Season During very low tides, it can be gathered in early summer to autumn. Try not to overharvest, and always leave the holdfast to produce new blades.

Edibility This is one of the more harvested seaweeds along the Pacific Northwest coast. The midrib can be cut into small pieces and eaten raw or, like the blade, dried and used for later use.

RECIPES

The recipes in this section are presented in the order in which their primary ingredients appear in this book.

Mashed Burdock Root

COMMON BURDOCK, *Arctium minus*, pg. 27

2 cups burdock root
1 russet potato
3 Tbsp butter
¼ cup (65 mL) milk
balsamic vinegar or lemon juice

Dig up (or purchase) three long burdock roots. Peel and cut them into 1 cm (½ in.) pieces. Dice a russet potato into similar sized pieces and boil with the burdock root for about 20 minutes or until soft.

Drain the water and add butter, allowing it to melt. Place in a blender with milk and blend until smooth. Serve on a plate with a splash of balsamic vinegar or freshly squeezed lemon juice.

Goes well with salmon and meat.

*Both common burdock and related greater burdock (*Artinum lappa*) have roots that are edible, though greater burdock roots are much larger. The roots from both plants should be harvested in their first year, before the second year, when they flower.*

Oregon Grape Juice

OREGON GRAPE, *Mahonia aquifolium*, pg. 145

1 kg (2.2 lb) freshly picked Oregon grape berries
1 kg (2.2 lb) dark seedless grapes
1 Tbsp (15 mL) honey

Put the Oregon grape berries and the seedless grapes through a juicer. Pour the combined juice in a pot and heat until just before boiling. Reduce heat and stir in the honey until it dissolves. Remove from heat and let it cool. Place in a container and let it sit overnight.

Makes approximately 1 litre (1 qt.).

As with any grape juice, this is also suitable as a starter for homemade wine.

Blueberry Pie

OVAL-LEAVED BLUEBERRY, *Vaccinium ovalifolium*, pg. 146

Pastry
2 cups (500 mL) all-purpose flour
¾ cup (190 mL) shortening
fresh cold water

Filling
1 Tbsp (15 mL) butter
1 lb (454 g) honey
½ tsp (2.5 mL) ground rosemary
1 Tbsp (5 mL) lemon juice
½ cup (125 mL) all-purpose flour
6 cups (1.5 L) local blueberries

Mix the flour and shortening together in a bowl until the flour-covered shortening forms small marble-sized balls. Gather in your hands and form into one large ball. If needed, sprinkle a little bit of cold water to help bind everything together. Cut the ball in half and flatten both halves into pancake-shaped portions. Put each portion into separate plastic sandwich bags and refrigerate for one hour.

Preheat oven to 200°C (400°F).

Mix and melt butter, honey, rosemary, and lemon juice in a small pot. Let cool.

Mix flour and blueberries in a separate bowl.

Roll out both portions of pastry with a floured rolling pin. The bottom portion of pastry should be larger than the intended pie pan. (I'm not good at carrying pastry without stretching it, so I fold it in four and then unfold it in the pan.) Fit the bottom portion of the pastry into the pan. Fill the pan with the blueberry mix, then pour in the melted butter mixture and fit the pastry roof on top. Be sure to cut a couple of slits in the pastry roof.

Cook for approximately 45 minutes or until you see the blueberry mix bubble out of the slits.

Licorice Fern Root and Trailing Blackberry Tea

LICORICE FERN, *Polypodium glycyrrhiza*, pg. 162
TRAILING BLACKBERRY, *Rubus ursinus*, pg. 155

approx. 25 cm (10 in.) of licorice fern root
a handful of trailing blackberry leaves
2 litres (2 qt.) water

Licorice ferns can usually be found on the coast in mixed forests on bigleaf maples. They epiphytically grow on the trunks with their roots covered in moss. When picking, only take short pieces of roots 2.5–5 cm (1–2 in.) long. Leave some roots in place to regenerate. Once picked, bring the roots home (or back to camp) and clean and cut them into 1.3 cm (½ inch) segments.

Put all ingredients in a pot and bring to a boil then let simmer for one hour. Let mix sit overnight. Next day strain off all solids. Put in fridge for cold drinks or reheat for a refreshing tea.

Bigleaf Maple Syrup with Apple-Blueberry Pancakes

BIGLEAF MAPLE, *Acer macrophyllum*, pg. 167
OVAL-LEAVED BLUEBERRY, *Vaccinium ovalifolium*, pg. 146

Pancake Mix

1 cup flour
1 Tbsp (15 mL) baking powder
1 cup (250 mL) milk
1 egg
2 Tbsp (30 mL) melted butter

Topping

several slices of butter
local apple, sliced
local blueberries
bigleaf maple syrup

Sift the flour and baking powder together then stir in milk, egg, and melted butter. Mix until smooth.

Butter a hot pan and place ¼ cup (65 mL) of pancake mix at a time. Cook until both sides are golden brown.

Stack pancakes with a thin slice of butter between each layer and top with sliced apple, blueberries, another slice of butter, and a generous helping of bigleaf maple syrup.

Though hard to find, bigleaf maple syrup is available in spring and early summer at selected markets.

Rowan Jelly

ROWAN, *Sorbus aucuparia*, pg. 183

1.4 kg (3 lb) rowan berries
2 tart apples
1 kg (2.2 lb) honey

Put the apples and berries in a large pot/saucepan with 2 litres (2 qt.) of fresh water and boil until the apples and berries are soft. Put the entire mix in a straining bag and leave overnight.

The next day, take the liquid and mix with honey and boil lightly for 20 minutes. Pour the mix into small sterilized jars and let set.

After setting, put the lids on and put in the fridge.

This recipe is more time consuming than complicated.

European Mountain Ash/Rowan berries have been used for over a thousand years in Scotland for making rowan jelly. It is supposed to be an excellent condiment with wild game. I tried to buy rowan jelly but could not find any locally, so I made some, and it's excellent with meat.

Golden Chanterelle with Salad and Risotto

GOLDEN CHANTERELLE, *Cantharellus formosus*, pg. 200

5–6 golden chanterelles
1 lemon
½ cup (125 mL) butter
1 clove of garlic, sliced
1 string of fresh pepper
1 sprig of rosemary
2 cups risotto
fresh greens for two servings of salad
10 baby tomatoes, halved

Clean and dry the chanterelles and cut each into 2 cm (¾ in.) pieces. Put in a bowl and drizzle with the juice of one lemon and let sit for an hour.

In a frying pan heat the butter, garlic, pepper, and rosemary. Bring to a light boil then reduce heat and let sit for an hour.

Prepare the risotto as per package directions.

Pan fry the chanterelles for approximately 10–12 minutes. Serve with risotto and salad, pouring the juice from the pan atop the risotto.

Makes dinner for two.

Pine Mushrooms with Salad and Potatoes

PINE MUSHROOM, *Tricholoma murrillianum*, pg. 204

4 or 5 pine mushrooms
1 lemon
½ cup (125 mL) of butter
1 clove of garlic, sliced
1 slice of ginger
pinch of thyme
10 baby potatoes of assorted colours, cut in half
baby greens for two servings of salad
avocado oil and balsamic vinegar to taste

Clean and dry the pine mushrooms and cut each into four or five slices. Put slices in a bowl and drizzle with the juice of one lemon and let sit for an hour.

In a frying pan melt butter on low heat with sliced garlic, ginger, and thyme. Heat until boiling then reduce heat and allow to simmer for one hour.

Place cut potatoes face down on a buttered cookie sheet and bake for 45 minutes at 160°C (325°F).

Put the pine mushrooms in the heated frying pan for 15 minutes or until golden brown.

Serve the mushrooms and potatoes, with baby greens.

Makes dinner for two.

Stir Fried Sea Asparagus

SEA ASPARAGUS, *Salicornia virginica*, pg. 217

3 Tbsp (45 mL) unsalted butter
2 sprigs fresh rosemary
1 clove garlic, sliced
225 g (½ lb) sea asparagus

Heat butter, rosemary, and garlic in a wok, but do not let it boil. Stir until the rosemary is covered in butter. Let sit/marinate for half an hour.

Completely dry the sea asparagus, and place it in the reheated wok. Stir the sea asparagus gently for 5–10 minutes then serve. Use the marinated butter in the wok as a dressing.

Sea asparagus is a fun little plant to serve with a main course or as a side dish. It gets people guessing. There are three secrets behind cooking it. First, use/pick only the top 10–15 cm (4–6 in.) portion of the plant—not only is this the tenderest, but it allows the plant to regenerate. Second, soak the stems several times in fresh water, even letting it soak overnight if you have the opportunity. Finally, do not overcook; keep the stems semi-crunchy, like lightly done asparagus.

ACKNOWLEDGEMENTS

THIS BOOK COULD NOT BE the quality it is without the help of many people, of whom I would like to acknowledge and thank the following: the University of British Columbia, with special thanks to the staff at the Beaty Biodiversity Museum—Karen Needham, Olivia Lee, Christopher M. Stinson, and Paul Kroeger; John Metras at the University of British Columbia; Warren Layberry, who edited the manuscript; and the great staff at Heritage House Publishing, including editorial director Lara Kordic, editorial coordinator Nandini Thaker, and designer Jacqui Thomas. And finally, thanks for all the support from the Pena family.

BIBLIOGRAPHY

Adolph, Val. *Tales of the Trees*. Delta, BC: Key Books, 2000.

Atkinson, Scott, and Fred Sharpe. *Wild Plants of the San Juan Islands*. Seattle, Washington: The Mountaineers, 1998.

Clark, Lewis J. *Wild Flowers of British Columbia*. Sydney, BC: Gray's Publishing Ltd., 1973.

Craighead, John J., Frank C. Craighead, Jr., and Ray J. Davis. *A Field Guide to Rocky Mountain Wildflowers*. Boston: Houghton Mifflin Co., 1963.

Harbo, Rick M. *Pacific Reef and Shore*. Madeira Park, BC: Harbour Publishing, 2006.

Haskin, Leslie L. *Wildflowers of the Pacific Coast*. New York: Dover Publications, 1977.

Hitchcock, C. Leo, Arthur Cronquist, Marion Ownbey, and J.W. Thompson. *Vascular Plants of the Pacific Northwest*. 5 vols. Seattle: University of Washington Press, 1955–69.

Lamb, Andy, and Bernard P. Hanby. *Marine Life of the Pacific Northwest*. Madeira Park, BC: Harbour Publishing, 2005.

Lyons, C.P. *Trees, Shrubs and Flowers to Know in British Columbia*. 1952. Reprint, Toronto: J.M. Dent and Sons, 1976.

Pojar, Jim, and Andy MacKinnon. *Plants of the Pacific Northwest Coast*. Vancouver, BC: Lone Pine Publishing, 1994.

Sargent, Charles Sprague. *Manual of the Trees of North America*. 2 vols. New York: Dover Publications, 1965. First published 1905 by Houghton Mifflin Co.

Sept, J. Duane. *The Beachcomber's Guide to Seashore Life in the Pacific Northwest*. Madeira Park, BC: Harbour Publishing, 1999.

———. *Common Mushrooms of the Northwest*. Sechelt, BC: Calypso Publishing, 2008.

Sheldon, Ian. *Seashore of British Columbia*. Edmonton, AB: Lone Pine Publishing, 1998.

Smith, Kathleen M., Nancy J. Anderson, and Katherine Beamish, eds. *Nature West Coast: A Study of Plants, Insects, Birds, Mammals and Marine Life as Seen in Lighthouse Park*. Victoria, BC: Sono Nis Press, 1988.

Stearn, T. William. *Stearn's Dictionary of Plant Names for Gardeners*. Portland, Oregon: Timber Press, 2002.

Stoltmann, Randy. *Hiking Guide to the Big Trees of Southwestern British Columbia*. Vancouver, BC: Western Canada Wildlife Committee, 1987.

Turner, Nancy J. *Food Plants of Coastal First Peoples*. Vancouver, BC: UBC Press, 2000.

———. *Plant Technology of First Peoples in British Columbia*. Vancouver, BC: UBC Press, 1998.

INDEX

Items in bold refer to the common names of species.

ABOUT THE AUTHOR

COLLIN VARNER is a horticulturalist/arboriculturist. He began his career at the University of British Columbia in 1977, working at the Botanical Garden, and for the next forty years he assumed responsibility for conserving trees in all corners of campus, including those planted by graduating classes and ceremonial trees dating back to 1919. Varner is the author of *The Flora and Fauna of Coastal British Columbia and the Pacific Northwest*; *Gardens of Vancouver* (with Christine Allen); and a series of popular field guides, including *Plants of Vancouver and the Lower Mainland*, *Plants of the Whistler Region*, *Plants of the West Coast Trail*, and *Plants of the Gulf and San Juan Islands and Southern Vancouver Island*. He lives in Vancouver.